Vintage Tomorrows

James H. Carrott and Brian David Johnson

O'REILLY®
Beijing • Cambridge • Farnham • Köln • Sebastopol • Tokyo

VINTAGE TOMORROWS

by James H. Carrott and Brian David Johnson

Printed in the United States of America.

Published by O'Reilly Media, Inc., 1005 Gravenstein Highway North, Sebastopol, CA 95472.

O'Reilly books may be purchased for educational, business, or sales promotional use. Online editions are also available for most titles (*http://my.safaribooksonline.com*). For more information, contact our corporate/institutional sales department: 800-998-9938 or *corporate@oreilly.com*.

Editors: Brian Jepson and Courtney Nash
Production Editor: Christopher Hearse
Copyeditor: O'Reilly Publishing Services
Proofreader: Christopher Hearse
Indexer: Judith McConville
Cover Designer: Edie Freedman
Interior Designer: David Futato
Illustrator: Rebecca Demarest

Cover photos courtesy of Jake von Slatt, who also built the devices pictured and who appears in the front cover image.

February 2013: First Edition

Revision History for the First Edition:

See *http://oreilly.com/catalog/errata.csp?isbn=9781449337995* for release details.

ISBN: 978-1-449-33799-5
[LSI]

James: To Amelia and Beatrix.

Both the future and the past are in your hands. Don't believe "can't." Take leaps of faith. Never choose an answer over a question. And make it better.

Brian: To James H. Carrott.

This book began with him and he has been the heart and soul of Vintage Tomorrows *throughout the entire project. Thank you James! You are a gentleman and a scholar and it's been a pleasure collaborating with you.*

Contents

Foreword: Any Questions?

Henry Jenkins

Vintage Tomorrows started as a conversation between a cultural historian and a futurist: what might steampunk (as a genre, movement, lifestyle, and philosophy) teach us about the ways people are thinking about their relationships with technology? They found it a gripping topic, they set out to talk with others, and the conversation kept broadening outward, as they found more thinking partners, many of whom were already talking through these questions together, and as one question led to another and another...

Christopher Columbus sailed west to get east; these authors looked into our collective fantasies about the past (or rather yesterday's tomorrows) in order to better understand our shared desires for the future. Ultimately, the authors hope people will design and build better tools because we have a deeper understanding of what makes technologies meaningful for people who spend a lot of time reimagining and retrofitting 19th century devices and gadgets. Makes sense to me.

I have been researching fan cultures for more than thirty years and participating in them for much longer, so let me put steampunk into a larger context. I am going to be painting with pretty broad strokes, so bear with me. Like the authors, I think something profound is going on here. *Vintage Tomorrows* compares it with the "counter-cultures" of the 1960s and I can follow them there. After all, as technology historian Fred Turner has suggested, the counterculture was the birthplace of today's cyberculture. For me, the explanation for all of this starts elsewhere—with the origins of science fiction and of science fiction fandom, both of which have in turn influenced contemporary forms of participatory culture.

The first thing you need to understand is that science fiction fandom has always been a community of people who were drawn together because they wanted to ask questions and most of the people around them found this constant probing to be annoying. The community's conversation started in the early part of the 20th century and hasn't slowed down since. Fandom doesn't simply exist to give fans a chance to talk with each other about the stories they enjoy. Science fiction stories exist to give fans something to talk about with each other. Speculation is the name of the game, and fans play that game better than anyone else.

The second thing you need to understand is that whatever steampunk is (and you will get lots of definitions here), it isn't "Victorian science fiction." If you asked Victorians about science fiction, they would not have a clue what you were talking about, though science fiction as a genre was profoundly and permanently shaped by the Victorian imagination. While many Victorians desperately wanted to believe in the stability of the empire and the power of tradition to give order and meaning to their lives, they were actually living at a moment of profound and prolonged change. One reason we are so preoccupied with the Late Victorian period, today, is that it may have been the last time, prior to the digital revolution, when our society confronted such a dramatic shift in its basic technological infrastructure. "New media" is a state of mind; every "old" media was "new" once, and people had to confront the changed realities that those new technologies might on the ways they saw the world. One way we cope with the shock of the new is to look backwards, hoping that the shock of the old might counteract its impact. So, the digital revolution has invited us to reconsider "The Victorian Internet," as Tom Standage described the Telegraph, and a range of other devices that look differently to us today thanks to our access to networked computers. Steampunk represents one of a number of new kinds of historical consciousness that are emerging as people try to put themselves in the places of previous generations as they underwent a media revolution.

Literary critic Jay Clayton's delightful book, *Charles Dickens in Cyberspace: The Afterlife of the Nineteenth Century in Postmodern Culture*, explores many different connections people have drawn between the Victorian age and our own. Contrary to common stereotypes about rigid Victorian thinking, Clayton argues that the late 19th century was characterized by "undisciplined culture": "[The 19th century thinkers] thrived in an atmosphere that might be described as predisciplinary, a world in which the professional characteristics of science as a discipline had not yet been codified...These men and women had an irreverent attitude towards boundaries and an impatience with anything resembling intellectual restraint. They mixed science, engineering, and the arts as they pleased. Without too much exaggeration, they might be described as the nineteenth-century equivalent of hackers."

The Victorians witnessed dramatic changes in the technologies of transportation and production, but for my purposes, what seems most interesting is that there were changes in technologies for communication and reproduction, starting with photography and the telegraph, extending to cinema and the telephone, and finally, hitting critical mass with the phonograph and radio. Many of these new technologies emerged with the fantasy of speaking with the dead: the phonograph was sold, in part, as a way of recording and preserving a loved one's dying words; the radio took shape in part because people believed that tapping into the "ether" might be a good way to contact the spirits; and photography took shape, in part, around the desire of parents to cling to fading memories of their children in an age of devastating infant mortality. Photography, French film theorist Andre Bazin, told us, was born with a "mummy complex." But, from the start, we used these new media to imagine the future in ever more tangible and concrete ways. So, go to the movies in the late 1900s and you would see magicians trying to imagine what it would be like to travel to the moon, or visit Coney Island and you might see Jules Verne's travel fantasies transformed into spectacular rides and attractions. This is how science fiction was born.

For science fiction to exist as a genre, the culture must experience such rapid change that people could recognize significant shifts over their own lifetime and thus begin to imagine a future that looks radically different from the present. A society where the same basic practices are handed down generation after generation has little use for science fiction. Another precondition may be the capacity of a people to recognize that things your society takes for granted are not the only "natural" or "logical" ways that people might live.

The expansion of the British (and other European) empires was bringing the western world into contact with what, for the Victorians, was an alarming amount of cultural diversity. The age was one that saw ongoing breakthroughs in geography (as people set out to map the empire), anthropology (as people discovered new people and practices), archeology (as ancient ruins were unearthed), and natural history (as the discovery of new species, such as the duck-billed platypus, shattered the conceptual frameworks by which people sought to order nature). The period was characterized by waves of immigration, which meant that western Europeans might no longer be able to confront these differences from a distance, but were dealing with other peoples on a day to day basis. Many early films sought to bring images of the far-flung empire to people back home, and the initial phantasies about television and the telephone had to do with enabling real-time connections with people on the other side of the planet. At the same time, people were haunted by the fear that there might soon be no uncharted territories left, that the frontier was closing, and with it, the safety valve functions frontiers provided for people who wanted to escape the societal constraints. Space, the future, the center of the Earth, all of it, represented the final (or at least next) frontier within the late 19th century imagination.

Most of the genres that would define popular fiction for the next hundred years trace their roots back to this period. Genres took on new importance as mechanisms for sorting products and identifying markets as culture was being produced and distributed on a mass industrial scale never before anticipated. But, this process took time. Themes about science, technology, the future, utopian societies, imaginary worlds, all flourished in the Victorian era, but science fiction in the 19th century was a genre without a name. Jules Verne thought he was writing adventure stories and romances; H.G. Wells thought he was writing social commentary; and some American writers turned to stories about mechanical men as an extension of the western. We can read these stories today as science fiction or at least proto-science fiction, but they would not have been understood as such by the people who first created and consumed them.

The Luxembourg-born, American-based magazine publisher, Hugo Gernsback gave science fiction a name (well, actually he called it "scientifiction") and a mission. Gernsback had been a leading figure in the amateur radio movement of the 1920s, at a time when this still-emerging technology was imagined as a community-based participatory channel (which might be deployed by scout troops, churches, and civic groups) rather than as a means of selling soap and soup or early on, getting people to come down to the department store and purchase consumer goods. The industrial revolution had profoundly changed the nature of technology; the basic tools people used were no longer something they jerry-rigged from the materials they found laying around and were becoming something that was produced on an assembly line and sold to consumers through mail-order or at the local department store (both great 19th century innovations in commerce). Science was shifting from something that emerged from the passions and interests of "gentleman amateurs" who shared their latest findings with the other members of their social clubs and becoming something which emerged within bureaucratic institutions (whether research universities or corporate think tanks). Knowledge was being disciplined and transformed from the stuff of salon conversation into the material of remote expertise.

The technological utopians, in the early 20th century, saw engineers and scientists as the new Philosopher Kings who should be ruling society with a firm but oh so very rational hand; Gernsback wanted to reintroduce democratic oversight into the equation, insisting that everyday people should have a chance to understand what was happening and to weigh in on the decisions that would transform their world. Gernsback wanted to empower his readers to tinker with these new technologies and speculate about the ways they would impact his society. Gernsback saw his publications as vehicles through which to foster greater science literacy. He began by publishing popular science magazines that reported on the latest findings from research labs or proposed hands-on projects that would allow his readers to play around with emerging technologies. Soon, he was publishing speculative fiction, alongside nonfiction stories, in the hopes of encouraging discussions that pushed beyond the limits of known science and anticipated the long-term consequences of technological developments.

Above all, Gernsback wanted to encourage his readers to ask questions about the nature of the universe (that "sense of wonder" old time SF fans like to talk about), the ethics of man's tampering with nature (a theme borrowed from romanticists like Mary Shelley), and the processes of technological change. Gernsback did something simple—he published letters in his magazines and encouraged his readers to get in touch with each other. Young readers around the country started to connect via the mails, then get together in person, and finally organize national conventions (ambitiously called World Cons), where they would engage in heated debates about the stories they read.

As they did so, they built upon an older infrastructure created in the mid-19th century by the Amateur Press Association, a group of mostly young men who published and circulated what we might now call zines. Many of the assumptions governing this community had in turn fed into the amateur radio movement, and indeed, many of them would feed, decades later, into the first online communities, since this group has consistently been among the earlier adopters and adapters of any new communication technology. These amateur publishers were experimenting with what it might mean to inhabit a community defined around shared interests without regard to geographic location, and they did so using the U.S. postal service. These science fiction fans were living within networked publics decades before such arrangements became widespread across the general population, and so, this niche community has lots to teach the rest of us about what it means to foster a more participatory culture.

Having constituted a community of active and inquiring readers, Gernsback, the publisher, needed to provide them with a steady stream of provocations and speculations. Gernsback initially dipped back into the Victorian era and began to re-publish stories from the previous generation which he felt might model what this genre might look like. Then, he reached out to pulp writers, who got paid by the word, to create new exemplars of this genre. Their stories were heavy on romance and adventure, light on scientific speculation, but for that reason, the fans enjoyed ripping them apart to see what might survive their exacting standards. Some fans wondered why Gernsback didn't signal where stories left known science behind, but Gernsback understood that debating the borders between hard science, softer speculations, and "crazy talk" was the driving purpose behind his forum. And before long, the first generation of fans, many of them still in their teens, became the next generation of writers, who embraced the virtues of "hard science fiction" as enabling more robust forms of speculation and debate. Almost every major writer in the 20th century emerged from this fan community and published their first works in its amateur publications.

These fans saw themselves as readers, thinkers, and potential writers, but also, often, as tinkerers and, again to use a somewhat anachronistic term, hackers, who could give material shape to the stuff of their imagination. We can trace this mixture of thinking and making back to Gernsback's original decision to run speculative fiction alongside popular and applied science stories, creating a point of intersection between geeks who hack technology and fans who hack stories. To be honest, early science fiction fandom was a boys' club—no girls allowed! There were always women who read and wrote science fiction, but they were not always welcomed by their male counterparts. Over time, as Helen Merick documents in *The Secret Feminist Cabal*, women transformed the established science fiction fandom, refusing to stay marginalized, and also adopted its practices to create an alternative space, more focused around media than literature, where they could write and share their own fantasies. If science fiction writers emerged from the ranks of science fiction fans, so did many scientists and engineers, who were drawn into their disciplines as a way of more fully realizing their fantasy technologies and so, science fiction has become a design platform that helps to set the agenda for future technological development.

From the start, science fiction has been a place where theory was consumed, with writers taking advanced concepts and pushing them to their limits, but it has also been a site of theory production, with writers and fans alike, encouraged to go places where no one has gone before. Andrew Ross's *Strange Weather* offers a good overview of some of the ideological struggles which defined science fiction fandom. H.G. Wells's readers might debate Fabian socialism and free love; Jules Vernes' readers might discuss how new transportation technologies were radically altering our sense of scale, allowing us to travel 20,000 leagues under the sea, from the

Earth to the Moon, or simply around the world in 80 days; Edward Bellamy's enthusiasts debated the merits of a technocratic restructuring of how goods were produced and exchanged in an industrial society, and Gernsback's readers wondered if radio might enable us to communicate with people from other worlds or other times. And this was only the start. By the 1950s, science fiction was helping people make sense of how Madison Avenue was using psychology and sociology to manipulate our desires. Science fiction was even giving rise to a new religion—scientology—which may or may not be an accomplishment, depending on your perspective. By the 1960s, feminists were using science fiction to debate and reimagine the relations between genders or how different kinds of sexual identities and desires might have begat alternative social structures. By the 1970s, science fiction fans were the ones who were learning Esperanto and supporting NASA. And by the 1980s, science fiction was giving us the language to make sense of cyberspace or debate genetic engineering, global warming, or the coming "singularity."

Steampunks, then, ask questions; they hack both culture and technology; they imagine alternatives, because that's what fans have always done. Steampunks returned to the Victorian roots of science fiction, before the genre even had a name, but then, they applied to those stories about lighter-than-air flight and mechanical men all of the protocols that had grown up, in the Gernsback era and afterwards, around hard science fiction. They looked at the flights of fancy from a hundred plus years ago, and asked what it would have taken to make them real. Historically, science fiction fans, for the most part, were asking questions about where we were going, and steampunks are increasingly asking questions about how we got here. The postmodern critic, Frederic Jameson, argues that science fiction is suffering a poverty of imagination, turning back on nostalgia because we've collectively painted ourselves into a corner and we don't know how to get out again. I suspect that's too simple.

We are at a moment when new technologies are altering our relationship with the past. A core fact about the modern condition is that we accrue stuff. Across the 20th century, we've seen dramatic increases in mass production and consumer capitalism: the average person, generation by generation, has access to more and more things. Yes, these things were not built to last; they were designed to be disposable, and the landfills that now cover half of our planet give testimony to how much stuff we've thrown away. But, as human beings, we've also sentimentalized our relationship to stuff. We hold onto old things that mean something to us, and almost any given item that is mass produced and distributed means something to someone.

One consequence of a networked society is that we now have platforms which allow us to exchange these goods with other people who would otherwise have tossed them into the trash: we are using eBay to re-assemble all of the toys and comics our mothers threw away when we went to college. Collectors used to gain prestige by having exclusive access to valued materials from the past; now, in the age of YouTube, they are gaining prestige by putting those media materials into broader circulation. People are selling off photographs of now forgotten ancestors to the highest bidder. And we are using webpages and blogs to share our expertise about materials from the past, and using discussion forums to connect with other people who share those enthusiasms, much as Gernsback fans used his letter columns to find others who shared their fascination with the future.

Critic Simon Reynolds talks about these phenomena as retromania and worries that our fixation with the pop culture past has started to crowd out innovation and experimentation. Yet, Reynolds notes that the conspicuous consumption of information, fueled by a network culture, means that many of the songs from the past are being heard for the first time, as people tire of the "standards" and go rooting around on b-sides and backlists. We might think of this as retro-consumerism, and smart companies are paying attention to older brands, for example, which develop cult followings, and starting to make and sell old products again. In many cases, the forces driving retro-consumption are relatively conservative: people who do not like the present are escaping into the past. In some cases, they are preservationists (thank goodness) and so people want the pure, raw, and unprocessed stuff. We are seeing early comics strips being re-printed in high end, over-sized books, and we are seeing people cut albums from old wax recordings; we are seeing people buy old tintypes and magic lantern slides; and we are seeing people gather together to watch performances which historically reconstruct the popular amusements of the late Victorian era. None of this is steampunk, though steampunk feeds on these same fascinations, and if you like one, there's some chance you will like the other.

While retro-consumption is typically nostalgic and restorative, steampunk culture is generative and reflexive: it wants to create a culture which never existed before, but might have; it wants to build on the material remains of the past and especially to tap into the outer limits of the Victorian imagination, but it also wants to create something unexpected from all of that brass, stained glass, and mahogany. Steampunks ask what would have happened if history had taken a different direction, what would have happened if Charles Babbage and Ada Lovelace had perfected the difference engine, if Captain Nemo and Phineas Fogg had really existed, if Tesla had been right, if the Hindenburg had not crashed and burned, if science had remained in the hands of quirky amateurs, if our culture had preserved a respect for beautiful craftsmanship rather than seeking to make cheaper and more disposable products for mass consumption. They are asking the same core questions that science fiction fans have always asked—what if, how come, and why not—but they are doing so by looking backwards into an imagined past rather than forwards into an imagined future. Of course, the Victorian age is only one such source for exploring yesterday's tomorrows: there are also people out there who really get excited by the "world of tomorrow" promised by the 1939 New York World's Fair and demand to know why they never got their jet packs; there are people who love the monumentalism of 1950s Socialist Realism, and wonder why all our futures can't be red; and there are probably people who are tapping into Latin American modernism, though so far, they haven't crossed my path. As Walt Disney told us once, "everybody's got a laughing place"!

And once the steampunks had started a conversation about what people of the 21st century might have to learn from their 19th century forebears, others started to ask awkward, unpleasant, and absolutely essential questions about whether we can really separate the things we find beautiful, quaint, and charming from the harsh material conditions which made them possible, from the other cultures which were being suppressed in the name of extending the empire, from the lives of the working class who worked impossibly long hours using insanely dangerous equipment so that others might enjoy a slower, more leisurely lifestyle. Can we have the age of steam without soot and smog? Can we have gentleman's clubs without gender and racial segregation? Who do steampunk fantasies speak to and for, and whose experiences, histories, and desires are being ignored?

Steampunks aren't just buying old stuff on eBay; they are building the kinds of alternative technologies they are imagining; they are creating objects that have the functionality of contemporary technology but have the aesthetic beauty and craftsmanship we associate with the Gilded Age. Every time they duplicate the function of new technologies using 19th century materials and processes, they demonstrate that other developments were possible. They are not simply trying to recover a lost culture; they are trying to imagine an alternative future that might emerge if we could only reboot history, if people in the past had made different choices, if people today had different values.

Steampunk's imagined history offers powerful resources through which we can challenge and critique contemporary realities. Steampunk is no more about the goggles than Cyberpunk was about the mirrorshades: they both simply constitute powerful metaphors for thinking about alternative ways of seeing the world.

Any questions? Good, then let's explore them together.

Prologue

"My mind wanders around, and I conceive of different things day and night. Like a science-fiction writer, I'm thinking, 'What if it were like this?'"

— CLAUDE SHANNON
(1948)

James Gleick begins his brilliant monograph *The Information: A History, A Theory, A Flood* (2011) with the story of a pivotal moment in contemporary history. In 1948, Bell Telephone Laboratories unveiled the transistor, "an amazingly simple device" that transformed our technological landscape, facilitating the waves of miniaturization that have made computing nearly ubiquitous today. At (roughly) the same time, working out of the same labs (more or less), Claude Shannon coined the term "bit," establishing a fundamental unit of measurement for information. Our mental world changed in 1948, and we knew it.

In the 65 years since, our media ecosystem has evolved into a world of ones and zeroes. The lines that so clearly differentiated one medium from another have become less substantial. A book, a film, an album, a game, even a web page... these are just the wrappers we put around information. Calling these things "wrappers" doesn't strip them of their meaning—far from it—but it's increasingly clear that the walls are coming down. Ideas, it seems, are the true currency of this contemporary age.

Why introduce this book like this? Because Vintage Tomorrows isn't just a book. It began with a single conversation and grew into an adventure that spanned the globe. Each question we asked spawned more, and we soon found ourselves enmeshed in a tale that spilled out over the rim of every container we tried to put it in. We think of Vintage Tomorrows as a project—a multi-faceted expedition through wild imaginary pasts, delving into the maelstrom of cultural change, searching all the while for the human face of technology and our dreams of a better future.

It seems almost trite to say it, but you can't put the future in a box. We learned that this is equally true of the past. Steampunk? Well, we'd hardly lead with the punch line even if we had one. For now it suffices to say that it is far more than the sum of its parts, and the currents that flow beneath its bubbly, be-goggled surface are broad and deep. We couldn't tell this story in a traditional manner. It literally defied our every attempt. So we gave in and let it lead us. We beg your forgiveness for the liberties we've had to take with linear time in order to tell this tale true.

You're holding the narrative heart of our story, but there is more. There's a documentary film, which we'll talk about more as we embark on our tale (if you like, you can cheat a bit and check out the trailer on the Vintage Tomorrows website (*http://vintagetomorrows.com*)). Also, a single book couldn't encompass what we learned. So we wrote another: *Steampunking Our Future: An Embedded Historian's Notebook*, to gather and share the fantastic material we couldn't fit in the pages that follows. It's free, and available for download (*http://oreilly.com/go/vintage*).

Welcome to Vintage Tomorrows! Please keep your hands and feet inside the ride at all times. We won't be using the brakes (we're not even sure there are any —never bothered to look), so hang on tight—you're in for a heck of a ride.

Acknowledgments

We owe so much thanks to so many that it was impossible to fit a full reckoning into print (under the reasonable assumption that our readers prefer a book that they are capable of physically lifting). As such, the brief acknowledgements below represent only the very tip of the iceberg. Please see the full acknowledgements online at the Vintage Tomorrows website (*http://vintagetomorrows.com*).

FROM JAMES

I have to begin by thanking my partners in crime, Brian David Johnson and Byrd McDonald. Brian, your faith in me and your endless enthusiasm for this project lit the fuse on the adventure of a lifetime—thank you for making this dream a reality. Byrd, your selfless passion and gentle dedication are an inspiration, and you made Vintage Tomorrows much more than the sum of its parts. Gentlemen, our collaboration has taught me more than I can possibly express. I treasure your friendship and doff my tarboosh in unending gratitude to you both.

I couldn't have asked for better editors than Courtney Nash and Brian Jepson. Courtney grokked from moment one and has been nothing short of a lighthouse on unfamiliar shores. Brian's sage perspective and unfailing wit provided a necessary anchor (it turns out we had just the right number of Brians). All the folks at O'Reilly Media have been wonderful—stepping outside their comfort zones time and again to help us create something truly unique.

In different ways at different times, Peter Rutkoff, Paul Boyer, Charles L. Cohen, and Ann Smart Martin took the risk of adopting this starry-eyed idealist into their tutelage. I owe them all debts that cannot be repaid (and no few apologies for bites to the hands that fed). My skill as a cultural historian is a product of their excellent mentoring. My failings and heresies are entirely my own.

Other essential thanks: Cherie Priest, Kevin Steil, Diana Pho, Libby Bulloff, and Claire Hummel for support and advocacy above and beyond the call; Chad Hessoun, Johanna Burton, Margaret Hogan, and Shelby Balik for being the right friends at the right time over and again; Shar Kunovsky for a set of tools unavailable in even the best stocked shed; Dave Cohen for essential audio wizardry; Katie Casey for amazing spreadsheet steampunkery; the Key West Literary Seminar for an outstandingly generous welcome; and finally Mark Thomson, Margaret Killjoy, and the crew of the Neverwas Haul for being themselves—true champions of the possible and role models for a better future.

Finally, I owe deep and special thanks to Margo Robb. While our partnership has changed in ways we never imagined, she has given me more over the years than I can ever hope to repay (and that without even mentioning our two amazing daughters). She has contributed to this book in ways I can't even hope to articulate. Our paths have parted but the past is always present. I cannot imagine having traveled these roads without you.

FROM BRIAN

This book and the entire Vintage Tomorrows project has been filled with some of the most amazing, intelligent, bizarre, and giving people you could ever meet. Without them coming along on this crazy idea that a historian and a futurist had this would have never happened. Thank you.

This project would have never happened if it hadn't been for Joe Zawadsky. I'd like to thank Justin Rattner for his support, open mind, and fearless exploration of new ideas. Much of the intellectual backbone of this book came from Dr. Genevieve Bell.

Finally I'm personally grateful to the big team that made this project so rich: Jim Olsen, April Miller, and the entire OMedia crew, Harlene Conley, John Metzger, Byrd McDonald, Alan Winston, Emmery Raw, Sherry Huss, Brian Jepson, Courtney Nash, Bruce Mau, Scooter Braun, Jamie Masada, Cory Doctorow, Bruce Sterling, China Mieville, Ken Hertz, Tim O'Reilly, Rich Bowels, Darrin Johnson, Sandy Winkleman, Shyama Helin, Doug Schick, Jeff Dean, John Croft, Antonio Tatum, Jake von Slatt, Sunny Jim and the whole Norwescon Crew, Donal Mosher, Mike Palmiere, Jay Melican, Tawny Schlieski, Karen Tanenbaum, Ashwini Asokan, Francoise Bourdonnec, Tricia Hinds, Shari Weisberg, Lama Nachman, Steve Brown Damon Sullivan, and Henry Jenkins. 25.

A Futurist and a Cultural Historian Walk into a Bar

It begins, like so many great ideas, over a beer. A futurist and a cultural historian have a pint in Seattle and start talking about the future and the past. They're both technologists, so it's hardly surprising when the conversation drifts into the topic of steampunk, a modern day mash-up of the future and the past, technology and culture.

The historian says that like the Beat Generation in the 1950s and the "hippies" in the 1960s, steampunk is a counterculture. It's a movement that gives us insight into how our culture as a whole is changing. Not only that, but it's telling us something specific about technology.

If you wanted to join the Beat scene, you had to know North Beach. If you wanted to hang out with hippies there was a song on the radio that told you to go San Francisco with a flower in your hair. But if you want to find steampunk today you need only look online.

Steampunks, along with maker culture, hacker groups, and any strange number of builder and creator networks are all tapping into the same underlying issue: our relationship with technology is changing. Even more interesting, we're aware of it—and for some reason we're rooting around in the past to help us sort it all out.

So over a beer historian James H. Carrott and futurist Brian David Johnson ask themselves: What can steampunk teach us about the future? What happens when we look backward in order to look forward?

James and Brian spent the next couple years traveling all over the world and talking to a wide range of people to get that answer. This is the story of what they learned, and what it means for the future.

It All Starts in a Bar

James H. Carrott and Brian David Johnson (Seattle, WA)

A FUTURIST AND A CULTURAL HISTORIAN walk into a bar... Yes, it sounds like the opening line of a joke but that's how it happened. It was cold and raining that night, which was really no surprise for November in the Pacific Northwest. Brian was in town for a lecture he was giving at the University of Washington and James makes his home in Seattle. We met at the Pike Brewing Company on 1st Avenue down by the mega-tourist attraction that is the Pike Place Market; it's the place where they throw the fish while tourists giggle and take lots of pictures on smartphones. There's also a big brass pig named Rachel—you're supposed to rub her nose for good luck. On that night the tourists had all cleared out to avoid the weather. Worked for us. We were there for the beer; Pike makes a great India Pale Ale.

The scene of the crime, Pike Place Brewery in Seattle, WA

We got together just to catch up. We'd met a while back at a tech industry show and discovered that we had a lot to talk about. It's always worth the price of a pint to walk away from a chat having learned something. This particular conversation moved lazily around until we started talking about steampunk. We were both fans of the fiction but Brian, ever the futurist, really hadn't spent much time thinking about it. It was all tied up in the past, right? We had been talking a lot over the last few years about digital culture and the kinds of futures that might come out of all this technological change. James had been exploring a connection between his counterculture research and steampunk, but had only just tied the future into his ideas. It's what happens when a historian and a futurist get together—they each start to look in the other direction.

Now we should pause here. If you have just said to yourself, "steam – what?" Don't worry. You are not alone. We'll spend plenty of time plumbing the depths of steampunk later. For now let's say that steampunks imagine what would have happened if we had 21st century technology in the 19th century. Jump back 150 years and spin around. It's like a science fiction of the past. If this makes you think about the 1954 Disney adaptation of *Twenty-Thousand Leagues Under the Sea* or the 1960 Rod Taylor movie *The Time Machine*, you're essentially in the right space. A more recent example is the 1999 film *Wild Wild West*, starring Will Smith and Kevin Kline—a fun popular film that wasn't a huge blockbuster but made up for that with nitroglycerine-powered penny-farthing bicycles, spring-loaded notebooks, bulletproof chainmail, flying machines, steam tanks, and a gigantic mechanical spider. Got a feel for it yet? If not, don't worry, you will.

Spotting Steampunk in the Wild

Steampunk spreads across many mediums: fiction, gaming, fashion, TV, film, art, you name it. Steampunks are into messing around with strange contraptions that look old but aren't. The easiest way to spot steampunk is by the aesthetic; the way it looks. It's all about gears and goggles, mad scientists gadgets, airships, and bustle skirts. It's historical, but weird... something is intentionally not quite right. Here's a few examples so you get the visual gist:

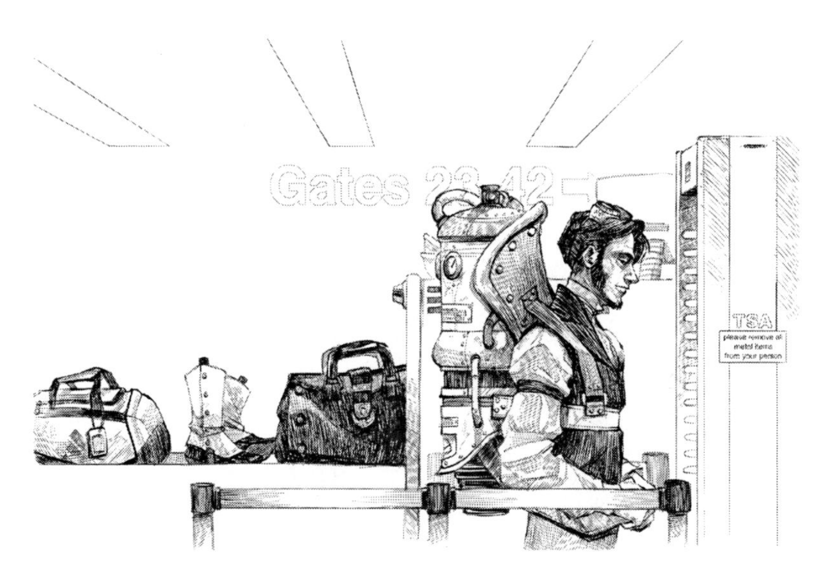

Claire Hummel, cover illustration for SteamPunk Magazine #+4

"The Queen of Cogs" (left) and "The First Joker" (right) from The Windrow-Ravenswood Deck (2010) (artwork courtesy of Dmitri Arbacauskas)

Paul Guinan, Boilerplate in the laboratory where it was created

Dashing desert explorer Willow Cooper-Barton at tea (photo courtesy of Andy Pischalnikoff)

"Victorian All-in-One PC" by Jake von Slatt (2008)

Now, steampunk's got a lot more going on than just goggles and gears —and we'll get to that in this book—but spotting it's the first step.

James took a sip of IPA and explained, "I've been mapping the cultural activity around steampunk."

"You've been mapping the cultural activity about steampunk," Brian replied, dead-pan, realizing that James had no clue how much of a culture geek he actually was.

"What?" James smiled, a bit defensively. "It's what I do."

"Go on."

"Ok," James continued. "If you think of steampunk as a counterculture, like the Beats in the 50s and the hippies in the 60s... If you look at steampunk like the natural progression from the Beats, the Merry Pranksters, and so forth, then something really interesting happens."

Note

James has a lot of background with the Beats and the hippies. The guy knows his counterculture—he has interviewed and studied such legends as Timothy Leary and Ken Kesey. He's spent a couple decades exploring subcultures and their effect on wider popular culture. More on that in our next chapter.

We'll hear the words "subculture" and "counterculture" a lot in this book. They may seem interchangeable, but there's a subtle, yet significant distinction between the two. A subculture is a group of people whose values or behavior set them apart from the cultural mainstream. A counterculture is a subculture that seeks to *change* mainstream culture. Fact: steampunk is a subculture. Hypothesis: steampunk is a counterculture. Keep an eye out for change.

"But for most people isn't steampunk really about girls in bustles with little hats and guys in goggles?" Brian asked.

"It's more like a cultural movement," James replied. "It's not just fiction and comics but also music and fashion and art."

"It's still about girls in little hats and guys in..." Brian insisted.

"I've starting working on a cultural heat map..." James interrupted.

Brian raised a well-arched eyebrow. "A cultural what?"

"Well, it's not actually technically a 'heat map' *per se,*" James demurred slightly.

Brian nodded, delivering his *I'm humoring you but only for the next couple seconds* smile. (Yes, he has a specific smile for that.) "James," he said, "skip to the punchline."

Note

You'll see this happens a lot when James and Brian talk. A cultural historian and a futurist don't always speak the same language. We also like to interrupt each other. And give each other a hard time. Makes for a fun pint.

James restarted, "It's kinda informal, but a cultural thermometer..."

"Now it's a thermometer." Brian's *not exactly sarcastic* tone.

"Whatever. The idea is to take in and map out all the cultural activity around a specific topic," James explained. "It could be a collection of activity around steampunk or it could be about the Super Bowl, doesn't matter. Ideally you gather up everything you can about the subject from media, fiction, art, fashion, movies, games, music... all of it."

"How much fashion activity do you really have around the Super Bowl?" Brian smiled.

"Jerseys, running shoes, whatever's in the coolest ad..."

"Good point."

James continued, "If you take the cultural activity for steampunk and plot it over time, you see this really interesting spike in activity starting around 2005 to 2007. There's this incredible rise in... well, steampunk stuff." (See next figure.)

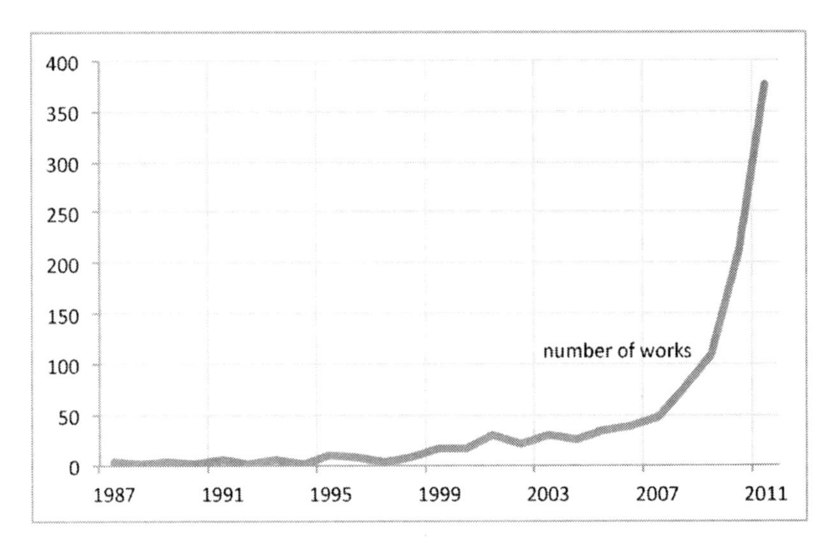

Growth in Steampunk Cultural Activity from 1987 to 2011

How Do You Read a "Steampunk Thermometer"?

If you're a smart, critical-thinking person (we're pretty darn sure you are—you're reading this book after all), you're probably asking yourself a couple good questions:

Where do these numbers come from? The chart you see above counts published works we found that fit our definition of steampunk (more on that later). What we mean by "published works" is: fiction, non-fiction, comics/graphic novels, TV, film, games, and music—the "hard goods" of cultural production. We're talking about stuff you could point at, find at your local library, or buy at a store. This is called "managing scope," which allows us to make our point without exploding our brains.

The numbers do not include conventions, performances, meetings, special events, clothing lines, works of art, local newspaper articles, Tweets, Facebook pages, websites, forums, organizations, businesses, and other things too weird to put in a bucket. Yes, steampunk shows up in all of these places. We couldn't count them all, but there are a lot of them.

Why did you limit yourself to these particular years? We didn't. There's steampunk as far back as 1947—again, more on that later—and we did count it. We just chose to start this chart in 1987 because it's the best vantage point from which to view the rise we're discussing. It's also the year that the term "steampunk" was coined. We stopped in 2011 because that's our last full year of data. Simple as that.

See the *Historian's Notebook* (see Prologue) for more details.

"That is interesting," Brian finished his beer and searched for the bartender. "Why do you think that is?"

James smiled and pulled out his iPhone. "I suspect it's got something to do with these things," he said, pointing at the smartphone. "Around 2005 these started coming to the market in mass. The iPhone came out in 2007. The Kindle, too. Android showed up the following year. All of these devices are changing our relationship with technology."

After ordering two more pints Brian asked, "Ok. That's interesting. Why?"

"I'm not sure yet," James replied. "But I think it's all about technology. Steampunk isn't anti-technology at all... it loves technology. First off, it tackles the implications of rapid technological change on history. Steampunks are imagining different technological pasts. But what I think is really interesting is that steampunk as a subculture exists *because* of technology."

Note

The iPhone was really the world's introduction to the "smartphone." Before that phones were pretty much just phones. They were dumb. Most people only made calls and sent short text messages. When Apple released the iPhone in the U.S. on June 29, 2007 the world was introduced to an entirely new experience. The phone wasn't just a phone anymore—it was a little computer and you could do some really awesome things with it.

Soon after, a wave of other little computers were announced and released. Amazon's Kindle followed on November 19, 2007 and Google released its alternative smartphone experience with Android in September, 2008. How we as people lived and interacted with computers had changed forever. We carried around powerful computers... in our pockets.

James's "Steampunk Thermometer" shows a distinct rise in steampunk cultural activity right around the same time that we witnessed a significant technology shift. James was definitely onto something.

"What do you mean?" Brian asked.

"If you wanted to be a Beat in the 50s you had to go to North Beach," James said. "If you wanted to get in on the *Summer of Love*, you had to go to the Haight-Ashbury. But today, if you want to be a steampunk, all you have to do is go online."

"Wow." This is where Brian really started to get it.

"And it keeps growing." James took a sip of his beer. "I think it's really telling us something about what's happening today in our broader culture. Just like the Beats and the hippies, steampunk is an indicator—a signpost telling us that something is going on, something is changing in our broader culture... and technology is right at the heart of it all."

The next thing that happened changed both of our lives.

"So if steampunk is changing the past," Brian asked. "If steampunk is designing and imagining a different technological past, then isn't it also making a very different request of the future? By playing with the past isn't steampunk designing a new future?"

We were both silent. We drank our beer as the rain fell outside on the empty Pike Place Market.

"That's a really good question," James said finally.

We could make up some crazy stuff about what we talked about next, but to be truthful, after that pause we just went on chatting and the rest of the night played

Auspicious pints

itself out in the kind of fairly regular way it does when two geeks get together and talk over beer. At the end of the night James went home to his kids and Brian went to the Watertown Hotel and prepared for his lecture the next day. Life went on as normal. But the question didn't die.

If steampunk is playing with the past, then isn't it making a different future?

When we were talking at the Pike Brewery in 2010, we were sitting at an interesting point in time. It felt like change was in the air—radical change. The way we were living with technology in our daily lives had changed pretty quickly all over the world. Technology had moved from the desktop to the laptop and then made its way into our pockets. Experts were projecting that by 2015 there would be 15 billion devices across the world that could connect to the Internet or to each other. That's pretty amazing. 15 billion devices by 2015. That's more devices than there are people on the planet! And what will be on those devices? Data. Lots of data. Google's Eric Schmidt had famously said that every two days we create as much information as we did from the dawn of civilization up until 2003.

It seemed like something was happening. Something was changing in how people used and interacted with computers and data. Steampunk seemed to be an area where these relationships and changes were playing out publicly and with grand flair.

Two days later Brian made a fateful phone call.

Sound of cell phone ringing.

"James Carrott. Hello?"

"James? James H. Carrott, it's Brian David Johnson.."

"Brian David Johnson! Hey man. What's happening? Where are..."

"I'm in my car driving back from Seattle," Brian yelled into the hands-free device clipped to his car's sun visor. "Listen. We have to do a research project."

"What?"

"What can steampunk teach us about the future? That's an awesome question. We need to do that! I don't know where or what it will tell us but we have to do it. I mean it. Steampunk! It's cool. We need to really look into what steampunk can teach us about the future."

"Right on," James replied. "You know I'm in."

And we were off and running.

From there, James and Brian took off and traveled around the globe—yes, we literally went around the whole Earth—sometimes as a team, but more often solo. Starting in Seattle we jumped up and down the west coast from Portland, Oregon to the big three in California: San Francisco, Los Angeles, and San Diego. Then we fanned out across the USA, to New York, and down to Florida. Then on to London, Ireland, France, Budapest, Hong Kong, Australia, and New Zealand, always to chase that elusive question: What can steampunk teach us about the future?

This book is the story of that journey. We started exploring steampunk culture, and that took us into the "literature of the future," the maker movement, cutting-edge design, and more. The first bit of the book is filled with the people we talked to, the places and things we saw and what we learned. (And then the documentary film crew showed up! More on that later!) We've talked to experts and gathered stories from what we saw along the way. And yes, we really learned a lot. We saw that steampunk could teach us something significant about the future. People were hacking culture and history, not just hardware and software. They were playing out a very different relationship with technology. But we didn't stop there. We asked: *So what does it mean? What do we build? What do we make? What comes next?*

We learned that people really do want a different relationship with their technology. They don't see technology as a cold dead thing that is cut off and separated from people. When you grow up with a smartphone in your pocket, technology is a part of your everyday life. Your devices are a part of who you are as a person. Technology, if designed correctly, can make us *more* human.

We learned that people want their technology to have a sense of humor, a sense of history, and most importantly a sense of humanity. These ideas are so simple but their effects are radical. The last bit of the book explores the implications of what we learned. We sought out and talked with a whole new crop of experts who, for the most part, knew very little about steampunk and the subcultures we had investigated. Well, at least they didn't *think* they did. These gracious people were nice enough to listen to our story and tell us what *they* thought it meant, and even went so far as to tell us what they saw happening in our culture and technology. The result was spooky. They were seeing and wrestling with the same issues we had teased out from our research. It showed all of us that we were on to something. Something was happening and it looked like a very different future than you might imagine.

So welcome to *Vintage Tomorrows*! Grab your goggles and your top hat, climb into the airship, and let's get going. It's going to be one fantabulous ride!

Beats, Pranksters, Hippies, Steampunks!

The Vintage Tomorrows project begins. Let's start with some history!

Our historian's path to understand the significance of steampunk, Makers, and hackers started many years before that fateful beer in Seattle. Throughout his career, James has lived his intellectual life on the fringes of our culture. His story takes us from the mud and gunpowder of a Civil War re-enactment battlefield, to a weekend with Timothy Leary, a chat "on the bus" with Ken Kesey, and finally to a defining moment in the Nevada desert at the Burning Man festival.

As James takes us through subcultures past and present, he introduces us to their luminaries and ultimately gives us a new view of the popular culture we thought we knew.

An Embedded Historian

James H. Carrott (Seattle, WA)

STEAMPUNK HAS a past of its own. Multiple pasts, actually. One of those pasts is in the realm of science fiction and fantasy; another is in the rebellion of contemporary counterculture. Turns out, I've spent quite a bit of time exploring both.

A few months after we walked into that bar, I met Brian in his hotel room in Seattle to review the research I'd conducted for him and Intel (at its earliest stages, the project was connected more directly with Brian's day job). We set a digital audio recorder on the table and Brian swatted away my books and my laptop.

"No, no, no. Bad academic," Brian said.

"I'm not an... academic." I insisted, clutching my *Steampunk Bible* (the most definitive overview to date) to my chest as he closed the lid on my laptop.

"Our book will not have footnotes," Brian insisted.

"No footnotes? What are we, barbarians?" I sighed. "Dammit, fine. I'll just make stuff up." I crammed the *Bible* back into my duffel.

"Good," Brian said. "That's when you're at your best."

Grumble...

"Dude, you're in this now. You're a steampunk. Let's start there," Brian pushed me to start talking.

"Wait, I'm still a historian." I insisted. "My methods are just a little, um... unusual."

Brian kept at it, "You're an embedded historian, then."

Brief introspective pause...

"You know what?" I stopped him. "You're right."

He was right. I'm an embedded historian: a participant-observer of the living past. I mulled the idea over while we had our after-working-session drinks, and it occurred to me that this has been part and parcel of my historical methodology from the start.

I Reject Your Reality and Create My Own

The best place to study change is from the fringe. Change doesn't happen *in* the mainstream, it happens *to* the mainstream. I can't claim credit for this idea. It really comes from Thomas Kuhn, and what he called "paradigm shifts." It sounds really heady, but the concept is quite simple. People build ways of thinking about things: frameworks that help them make sense of the world. These are "paradigms," the dominant way of understanding something. Like any structure—think of the framework of steel girders that holds up a building—they last until they can no longer withstand the pressures put on them (like damage from a fire, or even just slow rust and decay over time). When a paradigm can no longer stand up to the weight of anomalies, subversion, or other inexplicable things, it breaks down and needs to be updated, changed... *shifted.* In simpler terms, when our way of understanding things just can't explain it all away, we change the way we understand things. Humans are adaptable like that.

Where do those anomalies, subversions, and inexplicable things come from? Well, mostly from the edges of culture—the places where people think and act in ways that differ from the mainstream norm.

I started out on that edge. The child of divorced parents, I read too much and asked too many questions. While my mother and step-father were very supportive of my critical approach, I grew up during the Cold War, where such thinking wasn't generally encouraged. I'm not kidding about the Cold War: we had "duck and cover" drills in school in case of a Russian nuclear attack. (I'm not sure anyone thought that would actually save us.) I fully expected the world to end in my lifetime. I rarely trusted authority (see "divorced parents" and "read too much" above). It felt we were screwing up our world because apparently only *we* knew how things "should be" or "ought to be done." Darn those Russians, they're doing it wrong. Clearly we need a button to blow them up.

So at an early age I came to the conclusion that we were "doing it all wrong." Never one for half-measures, I couldn't help but question this stuff at its roots. I've also always been a firm believer that if something in front of you is broken, it's your responsibility to fix it. Credit where it's due, all of my parents taught me these values. My father always underlined a sense of duty, and my step-mother also provided an excellent example. (Yes, it's possible to be raised right in a "broken" home—there are two sides to every coin.) So I resolved to fix our culture. I didn't know it at the time, but this made me one of those subversive anomalies.

There's a catch, though. The paradigm has to shift *from,* but it also needs to shift *to* something. If you want to fix culture, you have to find models and alternatives. You can take it apart and re-fit the pieces, but the reason all those assumptions have staying power in the first place is the fact that people need structure; we need frameworks through which to understand the world and know how to act within it. Paradigms keep people from going insane in the face of Too Much World

To Understand. We start building frameworks as infants and never stop. This is something I learned over the course of my studies, both curricular and otherwise. The reason most well-meaning revolutions fail is that they don't have a well-functioning something good to replace the something bad they took down.

So you can't just take culture apart. That scares people. One great example: Dada, an early 20th century European avant-garde art movement that rejected reason and logic, and celebrated nonsense. (One of the movement's most famous members was Marcel Duchamp, who submitted a urinal as a piece of art for an exhibition in 1917—it was summarily rejected.) It's just what it sounds like, deliberate nonsense (hence the "baby-talk" name). Dada didn't last, and not just because "things change" or "those people are dead." It's because there really wasn't anywhere to go from there. You can deride and rip something apart until it's just a pile of scraps on the floor (or a urinal in an art show), but then what do you have? Once you've said "this culture is crap!" you're done. Unless you have something to put in its place.

In my early school years, I was a bit of a Dadaist. I struck out, but didn't build up. One year I flushed a towel down the toilet, which backed up the school's plumbing and shut the whole building down for a day. No budding Duchamp, I didn't sign the bowl, but my name was sewn on the towel. I even ended up in a cold war of my own as my parents and fourth grade teacher engaged in pitched battle over whether or not I could be included in the school's "gifted and talented" program. I was a bad student, my teacher argued, and a disruptive influence. "He's stifled," my mother and step-father argued, and presented IQ scores to the school board to prove it. Those few hours a week out of the regular classroom solving puzzles, playing strategy games, and thinking outside the box were a godsend. It's probably what kept me from getting expelled.

By high school, though, the public school system had pretty much beaten me into submission. I spent a lot of time in detention, doing the *Breakfast Club* thing. Pretty literally, too—I attended the same suburban Chicago school district where it was filmed. (For what it's worth, I'd probably place myself as something of a cross between Anthony Michael Hall and Ally Sheedy's characters—a slightly damaged, kinda awkward nerd with a bit of a chip on my shoulder.) And like the characters in this generation-defining, subversive movie, I'll be damned if I didn't find that detention held some of the answers I was looking for.

What changed? What answers did I find? Well, I found the punks and the (other) freaks. I began to see pockets of resistance and reasons for hope. I began to understand the mechanisms of my high school culture by paying attention to the folks who challenged it. I found people who were doing things differently; making their own rules. I saw that change was possible. I started to see frameworks that could take the place of the one I'd been struggling to break down. A punker friend introduced me to Doctor Who, bootlegging whole seasons for me on VHS. Dadaist no more, I had a universe of possibilities in front of me. Even the limitations of time and space are irrelevant to an imaginaut with a magic box that can take you anywhere, any*when*.

It was my American history teacher, however, who brought immediate and tangible change to my life. Dave Pasquini was an impassioned educator even in his free time, spending his weekends dressed as living history in Union blues. My eyes suddenly opened to a whole new way of exploring my questions about the world around me, and followed him enthusiastically into the weird and wonderful world of Civil War re-enactment. Dave died young, taken by a heart attack during my first year of graduate school. I like to think that he'd be proud of me and the path he helped set me on.

Living the Past

Re-enacting resonated particularly strongly with me because it gave me the opportunity to live in a different world. Yes, I was escaping my own world, but who doesn't need some escapism in high school? At least I didn't turn to binge drinking. It also gave me a way to fit cultural pieces together both through diving into the re-enactor subculture (a fascinating beast in its own right) and through living an imagined past. I did my best to leave the contemporary world behind and place myself in 1863. At least on weekends.

What did I learn? The past was dirty. It was smelly. It didn't taste all that great, either. I drank my first cup of coffee boiled to sludge in a tin cup, having bashed the beans with my rifle butt on a rock near the campfire. I drank the sludge. I needed it to get through the rest of the day. All this helped me realize the past is something *we* create. It's fundamentally human, with all the hairy, sweaty, and stinky bits. History is a manifestation of our own minds that comes from melding together what we learn from books with our own experience of what it means to be a human being.

My unit, the 20th Illinois Volunteer Infantry, was a rag-tag group of teachers, students, lawyers, plumbers, and even a playwright. None of us were set on glitz or

The 20th Illinois Volunteer Infantry, Summer 1990. The earnestly stern young corporal just to the right of our lieutenant in the front row is yours truly at 17. My teacher, Dave Pasquini is second from the right in the back row, with beard and slouch hat.

glory, and our entrancement with gear and garb was secondary to our desire to experience the harsh realities of life as Union soldiers during "The War Between the States." Our proficiency with fake blood and other special effects (I did a pretty darn good sucking lung wound) earned us the nickname "The dying 20th."

We never took ourselves too seriously (despite the look on my face above). History? serious. War? absolutely serious. Ourselves? well... not so much. Marching into battle one day, far from the waiting audience, my playwright friend and I shifted the column from *Battle Hymn of the Republic* (a well know patriotic song from the Civil War) to *Sunrise, Sunset* (an American show tune from 1971).

Just imagine 30 infantrymen wheeling lockstep into position, singing their way through the score of *Fiddler on the Roof.* We considered going on to *Yentl* (a Jewish musical movie from 1983), but Barbra Streisand seemed a bit too much. Looking back it looks like I was punking the past even at 16.

History Is a Living Story

> *We must not forget that the writing of history—however dryly it is done and however sincere the desire for objectivity—remains literature.*
>
> **— HERMANN HESSE**
>
> *The Glass Bead Game* (1943)

I've never really been able to get my head around the idea that anyone could think of history as "dead and gone," "behind us," or, heaven forbid, "dry and boring." I've read bad books and seen bad teachers. I know how to take the faceted and fascinating and turn it flat and dull. But that's not history. History is alive. At its core, history is about change.

Good historians don't close the book on their subjects, they open them. Though the same can't be said for all our sources (I've spelunked my share of county historical society basements), there's nothing dusty about the work historians do. The physical things—the books, objects in a museum, historical landmarks—are just bookmarks for your brain. History lives in your head. Each and every work of history is a work of living imagination.

Herman Hesse said that history is literature, and he was right. History is made up of stories—our stories—and those are things we live. Ask for advice and most great writers will tell you the same thing: write from experience—no matter how you play with them, draw your stories from who you are.

So for me to understand history and culture, I immerse myself in what remains and reach out to people to understand the patterns of history. To tell the story true I need to make my subject a part of who I am. Why? Because humans are humans, stories are stories, and those things make change happen. There's only so much you can learn from books.

After that, I couldn't see how I could write a paper about the experiences of Union soldiers at the siege of Vicksburg without putting on layers of heavy dark blue wool in 103 degree heat. I couldn't leave out packing 20 pounds of gear on my back, and facing down the sounds and smoke of enemy guns. I was still far removed from the actual horrors of the American Civil War, but being on a reenactment battlefield brought history to life for me. I can still recall the bitter, salty taste of gunpowder residue on my lips from biting into a cartridge. I can smell the tang of battlefield smoke and hear the ripples and roars of the guns.

Though I might not have said it this way then, I had come to the realization that history happens *now*. The stories historians tell are human stories, and they're as much about the time they're written in as the time they're written about. Like Herman Hesse said: it's literature. It's imagination and storytelling. We're missing something when we research and write at a remove. I admit to being a bit of a historical heretic, but I don't believe that we can really understand the people we write about unless we literally do our best to step into their shoes.

COUNTERCULTURE'S NOT DEAD, IT'S OUTSIDE LOOKING IN

Having salvaged my GPA, been surprised by a couple writing awards, and pulled off acceptable SAT scores, I made my way to Kenyon College in 1990, where I was quickly drawn to the fringe once again. I was fascinated by the artists, long-hairs, freaks, and weirdos. Even at a little liberal arts school in the farmlands of central Ohio, the "straights" held a good deal of sway, and man were they fun to subvert.

When I arrived on campus the first Gulf War was already in full swing. A few months later, I turned 18. I couldn't legally drink, but I had to register for selective service during a war. It was seriously scary.

It was painfully clear to 18-year-old me that the Gulf War was my generation's Vietnam. I'd spent enough time studying the horrors of war (take a long look at Matthew Brady's Civil War battlefield photographs and you'll see what I mean) to solidify a lifelong opposition to war in all its forms.

My head followed my heart, and I left the Civil War behind and engaged with the struggle against the current war. To me, this was as much about the past as about the present—everything was. In order to understand the 1990s opposition to the Gulf war, I had to look back at the Vietnam Era protests. This led me to 1960s counterculture. I discovered that it wasn't enough to simply oppose the war. You had to oppose the culture that made the war possible. My passion for counterculture and history hadn't just entwined, they'd fused.

Of course I wasn't satisfied merely watching from the sidelines. I read widely, but supplemented with a personal course of living research in 60s counterculture. By my senior year, I had a pretty darn good sense of what I was doing, and had begun to learn a lot—about myself as well as the past. On the first weekend of the new term, a friend and I took a couple hits of LSD and wandered off to explore the changing campus late at night.

Drugs and Living History

When I started to study counterculture, reading Allen Ginsberg, Jack Kerouac, Ken Kesey, Timothy Leary, and Tom Wolfe's amazing *The Electric Kool Aid Acid Test*, I realized that I could never really understand what they were doing without (so to speak) walking a mile in their minds. So in this case, living history meant some experimentation with psychoactive drugs.To be clear, I didn't enter into this lightly. I did extensive research before ever touching a single substance. I don't advocate this path to others, nor do I condemn it.

"Wow. This is like Ginsberg finding Greenwich Village and Kerouac and stuff," I rambled on. "You walk out your dorm room door and find this whole amazing new world of lights and people and..."

"Yeah, but look at that tree," my friend said. "That is seriously cool."

"Totally. It's life, roots and branches and history..."

"Sure..." she sighed.

"Timothy Leary tested psychedelics on divinity students," I said. I was reading a lot about Leary at the time. "Like 9 out of 10 of them had incredibly deep mystical experiences. This stuff can be life-changing in a seriously meaningful way. It's fun, you know, but it's so much more than that."

"Wait," she stopped me. "I just saw that name."

"What name?"

"Leary," she paused to think. "You did say Timothy Leary, right?"

"Yeah. He was one of the big acid gurus of the 60s," I replied. "He and Ken Kesey..."

"He's coming here," she blurted, a little weirded out by the synchronicity. The oddest things tend to come together in such altered contexts..

"No way!" My jaw dropped.

"Yeah way. We gotta go back to the dining hall," she pulled me back the way we had come. "There's a poster or something on the wall. He's totally still alive and stuff."

"I know he's alive," I said defensively, a little embarrassed I hadn't known he was coming.

Eventually we meandered back to the dining hall, moving from distraction to distraction.

"See?" she pointed at the poster on the wall.

"Wow!" She was right!

After I came down and got some sleep, I went back to the dining hall and confirmed that the poster was as real as it had seemed the night before. In my personal experience with psychedelics, the hype about "hallucinations" is pretty much just hype. No matter how stoned on what, my brain has never invented something out of nothing. Lots of twisting, changing, deepening, moving and such, even some synesthesia, but never hallucinations that just appear out of nowhere. Still, it never hurts to check.

It was true. I was about to have the opportunity of a lifetime. Historians don't generally get to meet their subjects (little did I know that this was only the beginning of my journey). Not only was Leary still alive, he was surfing the bleeding cultural edge, evangelizing the transformative power of visual communication and writing *Chaos and Cyberculture* (1994). Tim Leary had become a tech-head. And he was coming to my campus in a few weeks!

Through the folks who coordinated visiting lecturers and with the backing of the Department of History, I got myself set up to escort Dr. Leary around the campus during his visit, introduce him at his talk, and then give him a ride back to the airport. His lecture at Kenyon College was packed and I didn't mess up his introduction too badly (I was a bit overly-earnest), so I felt pretty good. Afterwards I escorted a 72-year-old Tim ("friends use first names") around the campus and we chatted about technology, change, and counterculture.

"What do you think your role was in the counterculture?" I asked. The campus bustled behind us—as much as a little liberal arts college in the middle of nowhere can bustle. "What does it continue to be?"

"Well there have been three countercultures," Tim replied. "There was the counterculture before: the people that drank booze and danced the Charleston. It allowed them social action in the '20s and '30s. Then you had the beatniks' counterculture of the '50s which was counter-Eisenhower."

"The '60s counterculture was different from..." he paused, clearly linking back to the broader story. "See, you have to compare the '60s counterculture with other countercultures. The countercultures a hundred years ago were made up of elite gentlemen who smoked opium and talked about Darwin. To be pro-Darwin was wildly counterculture about a hundred years ago. So when you talk about the '60s counterculture you have to compare it in the field of ongoing countercultures. The '60s was the first counterculture that was global because of radio, television, jet transportation, and records. So kids from Ohio could fly to Paris and be in a hostel and then hitchhike across to India.

"The beatnik movement was much more limited to North Beach of San Francisco and in New York. That's the tradition of coffee shop bohemianism. The hippie movement was simply a global explosion which was never possible before because the technology wasn't there.

"The '80s counterculture is high tech and the attacks by technophobic hippies on what I am doing now perfectly resembles the attacks from my mother's generation on Benny Goodman and swing on the radio and all that," Tim continued. "My role in the '60s counterculture pretty much is my role in the '80s counter culture. These waves are happening and I try to understand them and spread the word about it—to become a cheerleader for it."

So What the Heck Is a "Hippie"?

The word "hippie" was first used by North Beach bohemians as a pejorative to describe the young hangers-on who were starting to change the San Francisco counterculture scene in the early 1960s. Look it up on Wikipedia and you'll see a huge bucket full of people and ideas tossed together into a big ol' cultural salad. This is the way of the stereotype. We'll get into this more in a few pages when we talk about "beatniks."

So how did Tim go from being labeled a hippie himself to seeing them as reactionaries fighting against his work? Cultural change. By 1993, when Tim and I talked, "hippie" had mostly gone back to its derogatory roots, only this time it wasn't directed at young wanna-bes but rather crusty old "back to nature" types, mystics, and long-haired stoners. Some of these folks were at Tim's side on the cutting edge in the '60s. But Tim and others moved on into high-tech, while his "technophobes" sought a more pastoral future or were (justifiably) wary of technology as a tool that governments and corporations could use for surveillance and suppression.

This kind of thing happens time and again when words outgrow their britches. Watch for it with steampunk. As we'll see, it already includes a lot of divergent stuff. Even the biggest umbrella can only cover so many different ideas.

"Why do you think counterculture is so important?" I figured there was no better person to ask this question. "Why you are out here cheering for it?"

"Well if you don't have counterculture you end up with a monolith: a totalitarian monotheism," he replied. "Counterculture is the only way that we can evolve and that's why every established culture is against theories of evolution right now. The Catholic Church and Protestant churches certainly are telling people to hate the concept of evolution. The counterculture is almost always connected with media, poets, and writers and musicians. It always has to do with language and theater."

We'll continue to explore these connections throughout this book, but I want to call particular attention to that last bit. *Tim could easily have been talking about steampunk.* As we'll see, steampunk is all about art, language, and theater—messing around with culture to create new perspectives, many of which end up being pretty darn subversive. Counterculture both telegraphs and creates change. It also may well be the only way our culture can evolve.

"Where do drugs come into play with counterculture?" I asked trying to sound as studious as I could. My research had led me to suspect that drugs played a not-insignificant role in cultural change. "All the countercultures you've mentioned were involved in a lot of drug use. What do you think is the relationship between drugs and counterculture?"

"Drugs are always the key technology to a counterculture," Tim said matter-of-factly. He had, after all, been making this argument for many years. "Drugs fuck up your mind. They loosen you up and get you laughing, giggly or silly. Drugs get you dancing wildly and wiggling your hips and fucking all night. They also get you realizing with astonishment that things are much more complicated than the Boy Scout leader or the local minister made it sound. Anything that stimulates, shakes, or wobbles orthodox thinking is passionately opposed by the controlling culture. You simply can't have a rigid totalitarian orthodox culture with people that are getting wild and drunk. Intoxication has always been part of humanist religion because it's the way to not focus and escaping mind control."

I wondered about different kinds of intoxication, and I followed up carefully: "The '60s counterculture specifically had widespread use of psychedelics. I really think that's a new step." After all, getting drunk or high and goofing around was one thing, but my own explorations had been, at least as often as not, more thoughtful. "What do you think the specific effects of psychedelics as opposed to the alcohol-based countercultures were?"

"The wide scale use of psychedelic drugs in the '60s was because of the technology. It was jet planes and televisions and rock and roll records." Tim responded. "Before then you were limited to geography where the mushrooms grew or when the peyote button could be harvested. But with the wonders of Sandoz Laboratories' chemistry, suddenly people who lived in areas where there were no native psychedelics could get them. They were available to everybody, readily. That's a powerful technological step. We were no longer victims to geography."

"Before radio, northern white Americans never knew marijuana existed," Leary finished up. "The Mexicans did and then the blacks understood through radio and jazz music. Then the beatniks began to understand about marijuana. It was an enormous explosion of technology. After World War II it produced the '60s counterculture."

Not surprisingly, Tim was onto something big. The original intent of bohemianism changed so much with technology that the new bohemians weren't all that bohemian any more. As I returned from dropping Tim off at the airport, I mused about his lecture and our talk about the past and future of counterculture. Something different was happening. Change was in the air; new things were afoot. Maybe not the direct communication through light waves Tim specifically prophesied (well, not yet, anyway) but a shift in the technological paradigm. I'd made the connection between counterculture and technology.

That was 1993. Today, our technology has advanced to the point where we no longer need drugs to break our minds out of everyday reality. But don't think for a minute that things like portable computing and the internet aren't directly related to the psychedelic '60s. It's no accident that the San Francisco Bay area birthed the personal computer revolution. Many of the key figures behind the internet and the democratization of computing took psychedelic drugs and were a part of the '60s counterculture. An example: former Merry Prankster Stuart Brand took his *Whole Earth Catalog* online in 1985, creating "WELL" (the Whole Earth 'Lectronic Link)—among the very first online communities. Steve Jobs's 1970s experimentation with LSD ain't even the tip of the iceberg. Free your mind, and sometimes the rest really does follow.

We'll get deeper into counterculture's connections with technology later in the book. For now, though, Tim's given us a framework with which to understand the role of counterculture in social change, and sketched a pattern we can use to trace the evolution of counterculture itself.

Saying goodbye to Tim at the Columbus airport, September 1993 (photo courtesy of Sarah R. Kane)

We're still missing a piece, however, and that's the combined role of mass media and popular culture. For that, we have to learn...

HOW THE BEATS BECAME "BEATNIKS"

It's a dynamic as old as politics itself. If something gets popular enough, the folks in power co-opt it; they make it theirs and take it on as part of their agenda.

Subcultures and countercultures, groups that pose challenges to mainstream culture pop up all over the place all the time. A group of like-minded people get together and decide that they want to do something differently or think about something differently. Some ideas catch fire and others don't. The ones that do change things. They change themselves, and they change the culture around them.

The first contemporary (post World War II) counterculture to undergo this transformation was the Beat Generation. I began there with my thesis, and it still makes sense to begin there today.

In 1956, City Lights Press published Allen Ginsberg's *Howl and Other Poems.* The next year, Viking Press released Jack Kerouac's *On the Road.* In the course of a couple short years, a maelstrom of public attention swept their group of renegade writers into roles as spokesmen for a generation. Kerouac summed up the change this way:

> *Everywhere began to appear strange hepcats and even college kids went around using the terms I'd heard on Times Square in the early Forties, it was growing somehow. But when the publishers finally took a dare and published* On the Road *in 1957 it burst open, it mushroomed, everybody began yelling about a Beat Generation. I was being interviewed everywhere I went for* what I meant *by such a thing. People began to call themselves beatniks, beats, jazzniks, bopniks, bugniks, and finally I was called the* avatar *of all this.*
>
> **— JACK KEROUAC**
>
> *Playboy* (1959)

Before the national spotlight, the term "Beat Generation" was really only applied in the literary world to a small subculture of radical writers in New York and San Francisco. But in the wake of the publicity over *On the Road* it transformed into a label for a mass youth movement in opposition to mainstream culture. *Life* published an article titled, "The Only Rebellion Around," which described "the hairiest, scrawniest, and most discontented specimens of all time [are] the improbable rebels of the Beat Generation." Once they were labeled and stereotyped, the Beats' concepts of spiritual renewal and re-discovery could be readily dismissed as mere anti-society garbage, nothing new—"It's all been done."

When *San Francisco Chronicle* columnist Herb Caen mashed together "Beat" and "Sputnik" in 1958, coining the term "beatnik," the mass media had an easy handle: cultural rebellion was anti-American, smacking of Soviet Communism.

The term made it easy to group together Times Square denizens with San Francisco bohemians and anyone else who displayed similar countercultural ideas, dressed differently, or identified themselves outside the mainstream. The popular press represented these "beatniks" not as artists or spiritual leaders, but as a ridiculous puppet-rebel facade of shades, beards, and bongos—insane hipster-criminals who strove mainly to destroy American culture.

But in doing all this, the media shone a spotlight on the Beat rebellion, inadvertently reinforcing the counterculture's burgeoning sense of community. Popular condemnations of "beatnik" behavior fueled an underground mystique and attracted more curious young people to "Howl" and *On the Road*, where they could experience undiluted counterculture. The increased exposure of these ideas to the disillusioned youth of America proved to be one of the most important consequences of this media blitz.

"Beatnik" publicity also forced the counterculture to change. Any subsequent rebellion would have to alter its name and image in order to free itself from pastiche and cliché, but it turned out that even a vilified hip brought attention to some challenging new ideas. That attention changed the Beat Generation forever, but also amplified its challenge to a complacent postwar culture. The other thing it did was set the stage for '60s counterculture. The first "hippies," after all, were the young hipsters—the cool kids—who flocked to the North Beach Beat scene, drawn like moths to all that bright, shiny notoriety.

But was this all just a quirk of fate? Was it an accident that the Beats caught the eye of the mainstream media? Nope. Allen Ginsberg was all about getting attention—literally howling at the top of his lungs. People like Allen Ginsberg are often central to making the difference between a rebellion that fizzles out and one that catches fire. They're instigators and catalysts in our cultural fireplace—the match set to kindling, or the crucial gust of air that bursts it into a roaring fire. Ya gotta know the wood's dry, though, or you're wasting your match. The message is as important as the messenger.

But I digress a little... We'll meet some folks who may well be steampunk's Ginsbergs later, and we'll talk more about why its message resonates so strongly. For now, the point is that a small match in the right place can lead to a roaring flame. It burns ("ouch—darn it, now we're a freaking stereotype") but grows into something bright and transformative. The lesson of how the Beats became beatniks is that counterculture becoming hip and getting appropriated is part of the process of cultural change. That dancing little fire can end up changing the temperature in the whole room.

But enough metaphors and meta-talk—let's move on with the story. "Beatniks" may have appropriated the Beat Generation, but Neal Cassady, aka "Dean Moriarty," kept his show on the road (pun intended) and signed on with Ken Kesey and the Merry Pranksters. In keeping with the spirit of messing with history, our next step is a jump forward to look back.

STILL ON THE BUS

The 1990s were a rough time for counterculture. The Clinton presidency asked us all to pretend that one of our own was on the throne, and the cultural fringe lost a lot of its leaders.

Jerry Garcia died on August 9, 1995. I was moving that summer from my home in Chicago to start graduate school at the University of Wisconsin—Madison, but I skipped a day of packing to take my then 15-year-old sister to Soldier Field to see what turned out to be the Grateful Dead's second to last performance. Their show the following day was their last as the Dead. It could never be the same without Jerry. When Tim Leary died less than a year later, on May 31, 1996, it began to feel as if the dominoes were falling. The guiding lights of psychedelic resistance were winking out. And Allen Ginsberg, who passed on April 5, 1997 was barely a month in the ground when I met up with Furthur and its Intrepid crew of Merry Pranksters.

In May 1997, the Pranksters embarked on a promotional tour, bringing their (in)famous bus to the Rock & Roll Hall of Fame. Billed as the "Grand Furthur Tour," the affair was marketed as equal parts "last huzzah" and "retirement party." This is not to say that Kesey lacked for energy or that the bus was any less "The Bus" (of course it wasn't the original... that remained mired in the wetlands of Kesey's Oregon ranch)—his charisma met every expectation, and he handled the throng with sharp wit and twinkling eyes. As I wove my way through the crowd with my microphone and press pass, it was clear that this was no "final ride"—not for the Merry Pranksters, nor for the hundreds of people in attendance.

Preparing to meet people I'd been studying for years, I steeled myself to receive what I've come to think of as "the look." In my (then) limited experience, I'd found that people don't always take well to the idea of being interviewed by a historian. I guess it must feel a little strange to have a professional tell you that you're, well, "history." Maybe a little flattering, but also weird in that "but I'm still alive"-kind of way. It takes a delicate touch—one I hadn't really mastered back in 1997.

Furthur, "The Bus" in Chicago, May 1997

I needn't have worried. The Merry Pranksters didn't blink an eye. A wink or two, perhaps, but they got it. This was another element of the ongoing movie and having a historian interview them was no more or less weird than any other part of the adventure they'd been on for the past thirty-plus years. So I dug in and started asking about what was top of mind for all of us who were interested in cultural change.

I started by talking with George Walker. George, aka "Hardly Visible," was on the bus with Ken Kesey and the rest of the Pranksters as they made their historic voyage across America. That fateful journey, which included many "trips," signalled a massive change in our country's culture.

"There are people now who are saying that counterculture is over," I started off, reporter-style. The press pass wasn't just decoration, after all. I was also a volunteer reporter for WORT-FM, my community radio station in Madison, Wisconsin. "I've read that the hippies were something once, but they're old and dead now. Nothing new is happening in counterculture. Hip is dead. What do you say to that?"

"I say who says that?" George asked in return.

"There was an article in *The Nation* magazine last April," I answered. "It said 'hip is dead,' that it's all sold out and crusty."

"Well hip isn't any deader than God," he replied.

"Ha!" I laughed. "That's perfect."

"It's true," George explained. "We never got this many smiling faces around wanting to know what we were about in the '60s. There's more of y'all now than there was then."

"Were you hassled more then?" I asked. "I mean here the cops are just kind of keeping people off the streets..."

"We have almost achieved the status of being beyond hassle," he smiled. "You know we're like a circus. Everybody clears the road for the parade. Back then we were just a bunch of people out getting around, so we didn't really enjoy any status. We enjoy a certain kind of status now that I think keeps us not totally hassle free, but I think it really minimizes the hassles. We've managed to do it this long without anybody crashin' and burnin'. We've gained at least a little bit of respect. Even establishment people give us a certain amount of respect. Police give us a certain amount of respect, you know... It's different in that way. It's like we've earned our stripes or something over the decades. Most people realize that we're not monsters out there trying to destroy America."

He was right, of course. They'd changed culture. Shifted the paradigm. It wasn't all fixed, not by a long shot—and George knew it. He advocated constant vigilance—what the Pranksters had come to think of as a "warrior" stance. Just because things had changed didn't mean the battle was over.

My next conversation was with another of the original Prankster crew, Ken Babbs. I caught Babbs, a talented writer in his own right, seated behind a fold-up table at the end of a signing queue inside the Borders Books, signing copies of *Last Go Round*, which he co-wrote with Kesey. The book, a classic American tall tale spun around the early days of professional Rodeo (the 1911 Pendleton Round-Up), was Kesey's last novel. It's brilliantly punked Americana and a great reminder that playing around with history isn't just about steam.

"How is this different from what it was before?" I asked. "Right now you've got this whole audience put together by a big bookstore and everything..."

"Well actually it was put together by the Rock 'n Roll Hall of Fame," Babbs answered. "We didn't know anything about going to these bookstores until they gave us the schedule. It's all news to us. We've been getting a lot of reports from people who ask, 'Why are you going to Borders Books?' I just say, 'Well, they told us to. We do what we're told.'

"Well, obedience has always been a Prankster trait," I snarked.

Babbs chuckled. "Yes, we've always been very obedient. We follow the herd."

"How does this feel different from the original trip?" I asked.

"Well I don't know, that's a good question..." Babbs trailed off in thought.

"Now that Ginsberg's gone and Leary's gone and Garcia's gone, some people are saying that counterculture is dead," I said. "They say it's all over."

"I say look around you," Babbs replied. "It's not over. It's never over. You can't kill something like this. It's taken root now."

It sure had.

The next on my list was Carolyn Garcia, aka "Mountain Girl."

"I'm a history grad student," I told her, by way of introduction. Truth told, I was a bit in awe and wasn't entirely sure where to start, so I just blurted it out. "I write about counterculture in the '50s and '60s."

"Oh, goody!" she replied. That might sound sarcastic when written, but it felt honest when she spoke it.

"Where do you think counterculture is at now?" I asked. Yeah, I was recycling my questions, but I'd been getting such great answers. "I mean, Jerry passed on recently, Tim Leary is gone..."

"Counterculture is constantly changing," she answered. "What's prevailing today is this incredible corporate sales stuff. Everybody is wearing advertising on their clothes. We've all sort of accepted that as the way in America. At this point, counterculture is about individualizing your life, making your life as an individual and finding other individualists to club up with and gang up on the corporations. That's the essence of the counterculture today. Try to find our power base again as individuals and quit kowtowing to the corporately-managed financial spell that's been cast on this country."

"Can you define the central components to counterculture?" I pushed a little deeper. "Is it music? Is it drugs? Is it image? is it rebellion?"

"I think it's artistic expression," Mountain Girl replied. "It's free expression. It's artistic expression. It's exercise of liberty, which is very important, as we know. At the same time you embrace that which is good, and try to reject that which doesn't work or isn't working or is inherently bad. Each of us has to shuffle through life, work for the good, ignore or step past the bad, and give it no energy. That's still an important role for counterculture people."

This calls for a brief aside. She said these things to me fifteen years ago—in 1997—but it might as well have been yesterday. We'll talk about how steampunk plays into all this later, but unless you've been hiding under a rock for the past few

A chat with Ken Kesey, May 1997 (photo courtesy of Martin Alvarado)

years, you know just how relevant her comments about corporate culture, individualism, and artistic expression remain. Keep Mountain Girl's words in mind as we go on to Burning Man, steampunk, and beyond—'cause she had her fingers on the pulse.

Just when I thought it couldn't get any cooler, I finally I made it to the man himself: Ken Kesey, the Prankster-in-Chief. Kesey was the man who started it all. If the '60s counterculture was the Great Chicago Fire, Ken Kesey would be Mrs. O'Leary's cow. If it was the American Revolution, his 1962 novel *One Flew Over the Cuckoo's Nest* might well have been "the shot heard around the world." Okay, maybe that would have been *On the Road*, but Kesey's classic story of a mental institution turned inside-out certainly kickstarted the Merry Pranksters, and knocked over the psychedelic lantern that set the world ablaze.

Hardly knowing what else to say, I started off with the same question I'd been asking all day. "What do you think of the state of counterculture today?"

"I've never thought much about it," Kesey replied. "But since you ask, I realize *we* are the culture. All those people that are doing those dumb TV things and promoting hate...that's the 'counter'culture. People that are doing television and making cops heroes and making everybody feel like they need to own a Doberman to guard their place...that's the counterculture. Counterculture is tobacco. Counterculture is booze and armaments."

Yep, Ken Kesey's still in full form. He just flipped my question like a pancake, turning the world on its head in the process. I fumbled: "So, do you feel like you've won?"

"No, but I feel like if we don't," he replied sagely, "then we'll have to kinda start all over again in about 100 or 200 years."

He was a hard act to follow. "How do you feel now that Timothy Leary is gone and Alan Ginsberg is gone and Jerry Garcia is gone?" I asked slowly. "Do you feel alone? The Pranksters are sort of a voice in the darkness."

"No, not at all." A quick and certain reply. "For one thing I don't think we're finished with each other. I don't know anything about reincarnation but I feel like I'm still very much influenced by Leary and the way he died. Did you hear his last words to me?"

I never had. Still haven't seen or heard them anywhere else. I was enthralled. All I could do was shake my head in a mute "no."

"Tim's son called," Kesey continued, "and told me his dad was dying. I called back to talk to him and could just barely hear him. I said 'Well it's just been great Tim. Let's meet somewhere. Let's meet on Halloween night. You pick the place,' and he says 'Houdini's grave.' That's a nimble mind to be able to do that as you're dying. That's a warrior. This guy is teaching us how to behave. One of the mysteries that we all have to go through is the mystery of death, and dealing with it with humor, strength, courage, and a nimble mind."

Humor, strength, courage, and a nimble mind. Hold onto these ideas. We'll see them again (just wait 'til we meet Mark Thomson in Chapter 10).

Ken Kesey left us on my 29th birthday, November 10, 2001. I've never made it to Houdini's grave, but I'm sure he did exactly what he said he'd do. I don't profess any meaningful knowledge of the supernatural, but I do know that his spirit remains alive and well. Why? I found it nearly seven years later in the middle of Nevada's Black Rock Desert.

YOU GET THE BURN YOU NEED: DISCOVERING STEAMPUNK AT BURNING MAN

I was crawling through a festival way out west

I was thinking about love and the acid test

— JOE STRUMMER AND THE MESCALEROS

"Coma Girl" (2003)

Another flash forward, this time just over ten years, to Labor Day weekend in 2007. I'd left graduate school behind five years before, having decided that the publish-or-perish life of a junior professor was not for me. I loved my subject and my students, had a lot of fun in the classroom, but the sad truth was that a Ph.D. in history qualified you to do one thing: teach history. That would've been great if history professors were in great demand, but as I completed my research and started to write my dissertation, I looked around me and and found that the job offer from a little game company in Seattle held a lot more appeal than fighting other scholars for the scraps tossed onto the under-funded floor of the "ivory tower." I took a leap of faith.

Five years and two (fascinating and fun) jobs later, I had two kids, a mortgage, and was working for Microsoft as Global Product Manager for the Xbox 360 Console. Based on that little list, it might seem that I'd lost my subversive edge, but my interest in counterculture was as sharp as ever. When the offer came up, there was no way I could refuse joining my college roommate on his yearly pilgrimage to Burning Man, and of course, I was thinking about love and the Acid Test.

It's impossible to sum up Burning Man. It's a lot of things to a lot of people. It's a week of living and breathing art at a scale that's really unimaginable unless you've seen it with your own eyes. For some it's a giant party in the desert. For others it's a religious experience. For many it's where they go to feel most at home, to visit their true friends and become their true selves. There's really nothing else like it.

The basic facts: Burning Man is a week-long event in Nevada's Black Rock desert. It began in 1986 in San Francisco where a small group of friends gathered to burn a wooden effigy in celebration of the summer solstice. The event quickly outgrew its urban welcome, moved to the desert in 1990 and took off from there. The event is described by many participants as an experiment in community, art, radical self-expression, and radical self-reliance. Its focal point remains the same as it was in 1986—the burning of a colossal wooden man. And as you might suspect, it fed off that same old bed of counterculture coals—it's the descendant of Ginsberg and Kesey's fires. And man does it burn bright.

Burning Man 2012 (photo courtesy of Samuel Coniglio)

A Quick Burning Man Glossary

Black Rock City
: You don't just drop in on Burning Man, you live there. With a population of just over 52,000 people in 2012 (47,000 in 2007), for one week out of the year Black Rock City is one of the largest cities in Nevada—just about even with Carson City, the state capitol.

The Burn
: Shorthand for Burning Man, used for quick reference by those who've been.

Burners
: People who participate in Burning Man. The term is usually used to describe veterans or to identify kindred spirits.

Neon Night, Burning Man 2012 (photo courtesy of Samuel Coniglio)

The Man
: The physical and spiritual center of Burning Man—a huge effigy atop a structure designed with the explicit intent of burning in a beautifully staged manner.

The Playa
: The Black Rock Desert is a dry lake bed or, as it's called in the western US and Mexico, a playa. Its desolate surface is not sand, but a fine alkaline dust that gets into, onto, under, and between *everything*. "The Playa" capitalized refers to the specific place, in lower-case to the dust —one goes to the Playa and returns covered with playa.

I was excited, thrilled and well, just plain stoked as hell to attend the Burn. What that meant to me, though, is that I was going to *learn*. Being who I am (if you've met me, chances are you've seen me scribbling away in a notebook), I took furious notes so that I could recall as much of the experience as possible. Since there's no better way than to just tell it straight, here are some of those journal entries.

Wednesday 28 August 2007 4:45 PM

We arrived at Black Rock City around noon after a spectacularly long drive through miles and miles of desert scrub. The spirit of the place washed over me even as I got in line for my ticket at will call. Burning Man is the epitome of "come as you want to be." Anything goes (well, just about anything, e.g., you're no longer allowed to have sex or smoke pot in public), from spectacular costumes and elaborate theatrical camps to just plain ol' nudity. It was nothing short of astounding from moment one. As a Burn "virgin"—a newbie—I was initiated at the gate by a bunch of old naked hippies. I had to bang a gong loudly and yell: "I am a virgin no more!"

Black Rock City is hot and dusty, but it's also easily the most civilized place I've ever been. People here are living as they choose, adhering to a basic set of rules that provides for public safety, but otherwise making their own decisions about how to live their lives. This place is interesting to say the least. It's already made me think about human beings differently. The boundaries and expectations that govern our everyday lives just don't apply here. What better reminder that you can't truly understand some things without experiencing them?

Thursday 29 August 2007 11:25 AM

I woke to a mellow and comfortable morning. Actually got some sleep last night. Went out on "the town" until about 2 AM, and oh my lord—this place, combined, is the single most incredible thing I've ever seen in my life. Moving islands of neon roam through the desert. Huge transit buses converted into triple-decker mobile bars cruise from camp to camp. Platforms burst into flame in the sky. Pulsating geodesic rave domes cap the city's edges. Dozens upon dozens of strange events and gatherings all roll into and through one another in a blinking, vibrant, pulsating world. Not random at all, but chaos in its true fractal sense. It's another world, dedicated to passion, indulgence, creativity, fun, and art.

A woman just rode by on a bike, wearing a hot pink utility belt. My camp mates complimented her on her gear and she offered its contents: Chapstick and condoms, both pink. Her chosen role was as a mobile dispenser of these essentials. She was the Pink Protection Patrol. It seems like that's really the thing here. Think of something cool and thoughtful to share with other people and then open the door and share your heart with the human beings around you.

Thursday 29 August 2007 5:25 PM

We were just hit by a sandstorm. I am now one with the Playa. I have dust in places I didn't even knew I had. Ought to have expected as much, as my reaction was to dash right out into the blurring wind. It was absolutely magical wandering through a Mad Max wilderness, unclear even where the horizon was, just a blur of sand and sky and ground all melting into one big wall of spinning white/gray/brown dust.

I went out with goggles and a bandanna to protect my face (absolute essentials at the Burn for just this reason), T-shirt, shorts, and bag of water and simply

John Sarriugarte's Serpent Twins, Triobite, and Golden Mean kick up some dust, Burning Man 2012 (photo courtesy of Andy Pischalnikoff)

walked for hours through the sandstorm. I passed a bus converted into a giant pink birthday cake, just one among all manner of crazy vehicles that shuttled people across the Playa. Art cars of every variety roam through the streets—mobile bedrooms, tents on wheels, huge desert ships built up on the foundations of old buses, and giant cats. Imagination you can actually ride. There's even a huge robot ant building a structure out of giant mock sugar cubes. I think I've started to understand something: *the Burn is really about imagining the possible, making it real, and sharing it with others.*

Friday 30 August 2007 1:45 PM

On my way back from a brief foray to the porta-potties, I was mugged by a cute girl in nothing but frilly blue hotpants and a pith helmet (seriously, frilly hot pants and pith helmet!) who forcibly decorated me with pom-poms and cast-off torn-up TGI Friday's work shirts. It was my badge of shame for coming clad only in khaki. Guerilla costuming. Serious awesomeness!

Saturday 1 September 2007 3 AM?

WOW. In a word: WOW. Holy steampunk awesomeness! And that house! An entire Victorian mansion on wheels! Steam Engines! Gaslights! Penny-farthings! Amazing! Glorious Spectacle! Must sleep...

The Neverwas Haul on a glowing desert night, Burning Man 2012 (photo courtesy of Andy Pischalnikoff)

Saturday 1 September 2007 1:45 PM

I will never forget coming across that gaslit clearing at the Steampunk Treehouse, the huge crazy steam engine, and oh, man, that Victorian house on wheels. I'm in love! With a driving house! The heat woke me early and I had to go back and find it to make sure it was real and not just something the Playa somehow plucked out of my subconscious (as I've said, it never hurts to check). I tracked it down by heading to the treehouse and talking to the steampunks—easy to identify even in a land of goggles. They were the folks with big sideburns, waistcoats, turn-of-the-century trousers and bustles. They directed me to the house and told me that it's called *The Neverwas Haul.* It was parked on the Esplanade.

It wasn't *quite* as big as I remembered but it was no less cool. I went inside and looked around at the amazing attention to detail. There even was a framed *Harpers Weekly*-like cover black and white photo of the house with a bunch of people in the late 19th century period-weird costume. This crazy beautiful thing was clearly a creature of the imagination, but anchored perfectly in the past. I met the guy who made the structure—"The Major." I think I might have made a fool of myself by just blurting out, "This is *the single coolest thing* I've ever seen." He said, "'I'm so sorry to hear that.'"

The Neverwas Haul

Created in 2006 by Shannon & Kathy O'Hare, Kimric Smythe, and a crew of intrepid volunteers in six months on a budget that consisted essentially of pizza and beer, the Neverwas Haul has become a steampunk icon. It really is a three-story house, complete with parlor, viewing deck, and balcony—an epic-scale physical embodiment of the imagination. Just looking at the thing can flip a switch in your brain that makes the impossible possible.

In the six years since it first rumbled onto the Playa, the Neverwas Haul inspired countless steampunk makers and a veritable fleet of art cars. It's also a living thing, working only in symbiosis with its inspired, zany, and astoundingly coordinated crew. It may take a village to raise a child, but it takes a family to drive a house.

It suffices to say that the Neverwas Haul changed my life irrevocably and that I'm far from alone in this. When I returned to Burning Man in 2012 as part of its crew, I lost track of the number of people who came up to me and thanked me profusely for the life-altering inspiration they'd received from the Neverwas Haul.

Monday 3 September 2007 6:35 AM

Riding shotgun as we drive off the Playa, leaving civilization behind us as we return to the "default world." I've seen so much at Burning Man, but it's the Neverwas Haul that I'll carry with me forever. It's just so incredibly inspiring to see someone actually build something like that. I guess everybody who encounters this stuff for the first time must go through this kind of thing, but I'm just agog.

Back in college, Tim Leary suggested that I read William Gibson's books. Naturally, I started with *The Difference Engine*. I've known about steampunk for years, but I have never seen it on this scale. And real. Never outside of a book or a game. This is something new. Something is afoot and I need to know more.

What does it mean when people can build anything? What does it mean when they can let their imaginations run wild into the real world? Also, it's particularly interesting that when people let loose their imaginations, they often wander into the past.

THAT'S HOW IT ALL STARTED

On the Playa, I came face-to-face with full-frontal steampunk for the first time. And it blew my mind. Thing is, though, I'm still that kid who lost himself in Civil War re-enactments. I continue to be both "outside looking in" and neck deep in cultural history. That's a good part of what drew me to this new surge of steampunk. I saw imagination and the past stirred up together and made real, *lived in*, in a way that brought it home to my heart. This was living history as I had never seen it before.

Tim Leary, Ken Kesey, the beatniks, the hippies, and the Merry Pranksters led me to Burning Man—and it all flowed together so naturally. I study culture. I know a pattern when I see one. And this one was *glorious*. It quickly became clear to me that this wasn't just a personal revelation, it was the start of an intellectual voyage.

Something was going on. Steampunk *felt like counterculture*. It felt like cultural change in the making—what I'd been studying for years. But there was also something different about it—a new element that seemed connected with today's technology. Why was this happening *now*, when things like my iPhone and the Xbox 360 I worked on every day made me feel like I was living in the future? Was there a connection between steampunk's playful past and the cultural change I was seeing in my work as a technologist? I needed to know more.

The Major and his pilot, Professor Horatio Q. Birdbath, on the steps of the Haul, Burning Man 2012 (photo courtesy of Andy Pischalnikoff)

3

Technology That Ships Broken

Now that James has taken us through a tour of the countercultures and subcultures that led to steampunk, let's go talk to someone who knows a whole heck of a lot about it!

Cory Doctorow is an activist, pioneering blogger and technological commentator (Boing Boing), science fiction writer (*Little Brother*, *Makers*, and many more) and self-confessed massive fan of steampunk. Brian travels to London to chat with Cory about his take on the subculture and what it means for the future of technology. Cory always has an opinion and his passion and conviction is startlingly intense. He's been thinking and writing about these things for years. Brian also travels to London to talk with him about the deeper cultural and technological meaning that lies below the subculture and what it can tell us about our changing relationship with technology.

A George Orwell for the 21st Century

Brian David Johnson (London, England)

"YOU SEE there's a little feather on the second hand?" Cory pointed to the slender second hand as it made its way round the ornate little clock. "See the way the feather moves with each tick of the clock?" Cory held out his right hand and shook it limply each second. "See how it flutters with each second?" He continued to shake his hand but now he made a sound that's going to be nearly impossible to describe but I'll give it a try. The sound that a bestselling author and internationally known technology activist was making was a little like a turkey's warble—if the turkey had lips and was underwater.

I snorted.

"I know, it's really funny right?" Cory pointed at the little clock. "That's what I think steampunk is all about. All of it you can see in this clock. "

Detail of Roger Wood clock with the little feather on the second hand (photo courtesy of Porter Panther)

In the 21st century, few people have done more to popularize the thoughtful aspects of steampunk than Cory Doctorow. He's written and spoken about the

Cory Doctorow shows Brian David Johnson a ray gun

subject and the meaning of the subculture multiple times. He's written quite a few steampunk stories. But Cory is also a *fan* of steampunk. James talked a bit in the last chapter about "the Ginsbergs"—the people who fan the flames of cultural change. Well, Cory is one of those people, underlined and in capital letters.

"This one right here is my prize possession," he held up an ornate ray gun. When he handed it to me I could feel that it was impossibly heavy and lovingly made. "My wife gave it to me for my birthday. It's made by Greg Broadmore and the WETA team in New Zealand."

A Note from the Notebook: It's Not the Machine's Fault

"Love the machine, hate the factory" are the words of Margaret "Magpie" Killjoy, *SteamPunk Magazine*'s first editor and a shining light of steampunk's political potential. James and Byrd caught up with him at Gearcon 2011 in Portland, OR.

The complete interview is in our *Historian's Notebook* (*http://oreilly.com/go/vintage*), but we had to include Magpie's explanation of the slogan here. After this interview, our historian and director were pretty well convinced of Magpie's more-than-half-serious claim that "steampunk can save the world."

He explained that "*Love the machine, hate the factory* is actually a play off an animal rights slogan. It's about how we can actually interact with machines, essentially, as peers in a more neutral way. I'm not saying that they're animate. I'm not saying that it's alive, but everything represents a debt to the earth: everything that has been pulled up out of it should be given weight and importance.

"What I'm saying is it's not the machine's fault that they have been tied down in factories and been used to make the machinery of war or the machinery of social alienation—which is, essentially, what I think wage labor is—but the concepts of the machines, themselves, are not to blame. And I don't think that we should destroy all of the machines. I think that we should destroy industrialization.

"I would say that, right now, technology is essentially profit-motivated. And I would argue that that approach actually does a disservice to science. The technological lines that we develop along now are not really the most interesting lines. I believe that we will be forced to reconsider all of the motivations behind our technology by ecological concerns. And I think, as we do that, we're going to start coming to some of the same answers that steampunks are on the forefront of. Well, except for the ray guns."

We all wondered whether Greg Broadmore would agree with that last bit. James had a few beers with the famed ray gun magnate in New Zealand. That rollicking interview is also in the *Historian's Notebook* (see Prologue). Read it and find out.

I had come to spend the day at Cory's office in the Hackney Borough of London to talk about steampunk and this idea that James and I had that it might just be telling us something important about the future. Why would this matter to me? I'm the resident futurist at The Intel Corporation, the company that makes the chips and intelligence that goes into your computer. It's my job to look ten to fifteen years into the future and figure out how people will act and interact with technology. This might sound a little bit like science fiction but it's actually pretty pragmatic. It takes about five to ten years to design and manufacture the chips, so it's of vital business importance today for Intel to know what people will want to do a decade from now.

To come up with this vision for the future I use a mix of social science, technical research, economic data, interviews and even some science fiction to model what it will feel like to be a human in the future. In the work that I do one of the things that I'm beginning to see is that people's relationship with technology is changing. Because of this I'm constantly hunting for new ways to look at how people are using technology. That is what led me to steampunk, and what brought me to London today.

It was a cold clear day as I walked to Cory's office. Hackney is a working neighborhood. You won't run into a lot of tourists, just a lot of folks going about their days, getting stuff done. It's also home to Shoreditch, where Cory's office is located. It's a bleedingly hip and sometimes clichéd neighborhood, like San Francisco's Mission District or Brooklyn's Park Slope.

I have to admit, it took me a little while to find Cory's office building. I didn't really know what to expect. I'd been working with Cory for some time but I'd never been to his office. The notoriously transparent blogger and writer chronicles his life, fiction, and activism on a daily basis. If you ask people to define Cory they are somewhat stumped. Is he a science fiction writer? Sure; he's written seven books and dozens of short stories and he's still going strong. But that doesn't quite nail it. Is he a journalist? Yeah, he's a prolific writer for publications like the *Guardian* and *Publishers Weekly*, but that's not all of it as well. Is he a blogger? Well, yes, that too. Cory has defined an entire generation of bloggers as the co-editor of Boing Boing (*http://boingboing.net*), his weekly podcasts (*http://craphound.com/?cat=6*), his prolific Twitter feed (@doctorow), and wildly personal and intriguing Flickr posts (*http://www.flickr.com/photos/doctorow*). That might get close to it, but it doesn't capture who he is. Is he an activist? Certainly! Cory has made his name by vocally and vehemently fighting for personal rights in the often confusing and confounding digital age. Ok... then so what is he?

"He's our George Orwell of the 21st century," someone once told me. That's a heady comparison. But to most people who know him, Cory Doctorow is just, well, Cory.

So where would the George Orwell of the 21st century have his office? Where would he set his command center? I passed a garage with tattooed guys banging away on a snarl of questionable cars. There were the boarded up buildings and across the street, the entire block was taken up with a construction site. Overhead the train ran every few minutes, with its clang and ear splitting squeal. It's at these moments when you realize that mass transit is a railroad, a loud and heavy overland passenger rail with massive cars that scrape across metal tracks. And how appropriate for thinking about steampunk.

There was no way of avoiding it. I was lost. To the mobile phone!

Brian David Johnson lost in the streets of London looking for Cory (photo courtesy of Porter Panther)

Cory talked me in and I found the building. I'd actually walked right past it multiple times. I didn't even know it was an office building. It looked more like a parking lot with a squat storage facility attached. I made my way up the stairs to the second floor that smelled a little like most parking lot stairwells in urban cities. Yes, I mean it smelled of urine. Through the heavy security door and down the dim hallway I was confronted with a long line of heavy red doors. Now from the signs on the walls it was obvious that there were businesses there. This made me feel a little better. There were some tailors, a few software shops and a couple of spicy magazine/journal publishers. But a couple of the doors had a heavy padlock securing them. An old-fashioned padlock! I wasn't sure people really even used padlocks anymore, but there it was.

Down the hall I found "NONDESCRIPT BORING HOLDINGS," knocked on the door, and out popped Cory. With a smile he led me into his office and the world completely transformed into a geek wonderland.

The first thing I noticed as I walked into the office was that the place was crammed with books. Two walls of the office were lined nearly floor-to-ceiling with them, and on the other side were two massive book cases packed solid. Some of the books are Cory's translations, reprints, and his uniquely handcrafted special editions. Cory is a legend in the publishing business for challenging just about every preconception of how to publish and sell books. He gives away all of his writing for free online via Creative Commons (*http://creativecommons.org*). Yet he is still a *New York Times* best-selling author, confounding many people's notions for how to write, market, and sell books in the information age.

But Cory didn't stop there. He became his own publisher, creating hand-bound bespoke editions of his work that include everything from one-of-a-kind end leafs to digital editions on inset SD cards with embedded pieces from Cory's creative process. All of these books were there, too. This is the center of Doctorow's business world; where he creates, publishes, ships, and broadcasts everything he does.

Contraptions, Coffee, and Culture

The rest of the office is filled with kit, kitsch, and creature comforts. I really mean "the rest of the office"—every nook and cranny. Cory makes a perfect cup of coffee and is rather obsessive about it. He's something of a mad scientist when it comes to coffee, with the bizarre apparatuses to back it up. Even his coffee is steampunk-inspired: one of his machines is a cold drip contraption that stands four feet high and takes hours to brew.

Coffee in hand we sat down to chat. To start things off I asked for his general description of steampunk. It's a contentious question for many people and something we'll continue to explore throughout this book. You'd be surprised how many blogs and panels have spent quite a lot of time arguing this very question. So naturally that's exactly what I wanted to begin with.

"I'm skeptical of definitions," Cory answered. "Especially definitional questions about subculture. The problem is that they usually lead to fruitless arguments like: Are you punk or not? Are you Goth or not?"

"But by way of introduction I'd say it's stories, movies, pictures, and objects that all seem to come from an alternate history," Cory explained. "It's a different past where people could make incredibly polished and amazing products but these objects don't come from a factory. People make these things on their own. Steampunk is full of mad inventors building incredible fantastic contraptions," Cory waved his hands and pointed to the various objects in his office. "Some of which seem to have parts of our contemporary technology in them. Sometimes these technologies are computers, but they're made out of gears and powered by steam."

"Steampunk is really an alternative history that's set in a nebulous period sometime between the Victorian era and the 1930s," he finished up. "It's full of grand and glorious, stylish people doing beautiful hand-tooled work with technology that, today, we would associate more with an assembly line in Shenzhen, China."

Knowing that Cory was a long-time fan of steampunk, I wanted to get an idea of how he came to it personally. So I asked.

"I was working in a science fiction bookstore when *The Difference Engine* came out," he replied. "At that time I was paying close attention to everything in the field. The cyberpunk novels were very big. William Gibson and Bruce Sterling (we talk with Bill and Bruce in Chapter 9 and Chapter 17, respectively) were great, towering giants of the field. In 1990 they got together and published this strange collaborative novel.

"There had been other steampunk books before then. There'd been novels by James Blaylock, Tim Powers, and K. W. Jeter," Cory continued. The more I talked with Cory the more his encyclopedic knowledge of all things steampunk and science fiction became obvious.

"Jeter had actually called it *steampunk* in the pages of *Locus* magazine." *Locus* is a rough-edged alternative art and lit mag. Cory's read all this stuff, but most of us would never even stumble across it. "But it wasn't really the kind of thing that the average person who came into the science fiction book store said, *Have you got the new K. W. Jeter steampunk novel?* It was still too obscure."

"The book that got everyone paying attention to steampunk was *The Difference Engine,*" Cory paused and smiled. "Now when I say *everyone,* I do mean a very small group of everyone. I really mean within the mainstream of hardcore science fiction readers. All of a sudden, all those people knew what steampunk was. But after that it pretty much fizzled. After *The Difference Engine* there was the occasional novel that came out, but no one was really paying any attention, at least in the English-speaking world.

"Then next time I encountered steampunk was in the mid part of the 2000s. Eight years later people started to send me bits and pieces of things they had made or written for me to post on *Boing Boing*. That was really amazing! There were all these costumes, and gadgets... especially gadgets... tons and tons of gadgets and sculptures that were being made in the steampunk style."

"At the time, I thought it was really weird." Cory smiled again. "There weren't really any other steampunk stories out there. There had been *Wild Wild West* and a few other movies, but it wasn't like it was everywhere, the way cyberpunk had been in its day."

Cyberpunk

Steampunk and cyberpunk were 1980s science fiction's fraternal twins—born at the same time, but looking in different directions. When Jeter coined the name "steampunk" with his tongue planted firmly in his cheek, he was riffing off of cyberpunk. Cyberpunk was a hot new subgenre of science fiction that explored the newly connected world (at the time) of computers and the Internet.

Today, cyberpunk is almost as much science fact as science fiction. It still shows up in places, like *The Matrix* trilogy but a great deal of its aesthetic and a lot of its ideas about cyberspace and enhanced reality are just part of our daily lives in the 21st century. This fact leaves some to argue that the proliferation of personal computers, smartphones and tablets as well as the world's daily reliance on the Internet has made us all cyberpunks. Now cyberpunk is the new normal—which might have something to do with the rise of its past-punking sibling...

"Then in 2006 I was one of the guests of honor at the French National Science Fiction Convention in Nantes," Cory recalled. "It was filled with people dressed up in steampunk gear and they were all reading steampunk comic books. That's when I made the connection that steampunk is best as a visual style. It shines when it's visual."

Note

As you may have already caught on, Cory rarely misses a cultural beat, and it's worth pointing out that he hit an important nail on the head. Something poured new fuel on steampunk's fire in the mid 2000s. That something had a lot to do with making—physically building things, not just writing about them—and technology. It's what James saw at Burning Man (in Chapter 2), and also the reason that I'm here. Throughout our travels and conversations, the same few years (2005-2007) keep coming up as moments of critical connection. 2006 was the year that the Neverwas Haul first rumbled onto the Playa at Burning Man.

"French graphic novels have always had a greater adult readership than American comics," Cory explained. "It's always been a much more respectable medium and had a greater variety of subjects and titles being published. So there in France at the French National Science Fiction Convention in Nantes, all of a sudden I realized, *Oh wait, this is where steampunk went. It went to France and became a comic book genre.*

"It really seemed to take off from there. The gadgets became really successful and popular on the Internet. There were costumes and more comics, and even some movies. Then all of a sudden there were books again. Cherie Priest came out with *BoneShaker* and Scott Westerfeld came out with his *Leviathan* series (we'll spend some time with Cherie in Chapter 4 and Scott in Chapter 7). They both started doing some really great work in the steampunk genre.

"What was really interesting was that some writers started doing young adult fiction," Cory said. Cory has a particular interest in young adult fiction. His 2007 young adult novel *Little Brother* (making reference to Orwell's Big Brother from *1984*) reached the *New York Times* Best Seller List). "This meant steampunk started taking off with a whole new generation of readers that really had never heard of William Gibson, Bruce Sterling or *The Difference Engine.* You can think of steampunk as being born as a literary genre that fizzled out and then sizzled underground for a while, then became a comic book genre in France, and then that spurred a whole new literary genre and subculture."

Note

Steampunk also stayed alive in the U.S. in the realm of tabletop strategy and roleplaying games. A fringe audience, to be sure, but a loyal one. Cory's absolutely right about the vitality of steampunk in French comics, but when culture is concerned, there's always a little more than meets any given eye. This is also a good demonstration of James's half-serious theory that it's the freaks and the French who sharpen culture's cutting edge.

Bodging About Over Lunch

The day was still lovely, so Cory took me to a café within walking distance of his office. The café was located in an old shed within the walls of an enormous Victorian school. The school still functions, ringed by what had once been some of the most notorious slums in east London. Now the neighborhood is filled with much sought-after executive flats.

We arrived just as they opened, and sat by the open doors. The restaurant's owner made his name with another venture called St John's, which specializes in "nose to tail" dining—that is, all the cuts you're usually used to plus offal and guts, cooked with a lot of unapologetic gusto. The menu in the shed was a smaller version of this, with a lot of traditional English hedgerow vegetables and tureens made from innards. "Punked" pigs seemed somehow appropriate.

On the walk over we talked about how steampunk appeals to so many people's inner engineer. It's about building and making and hacking. This is what makes steampunk so interesting to a great number of its devotees. But steampunk also seems to have a strong point of view, and I really wanted to know if steampunk also appealed to the activist in Cory.

"Steampunk ships broken. That's what I like about it," he said as we crossed the street. "It's comes to you unfinished. A steampunk object arrives with the intention that it's not meant to be used when it comes out of the package. You're meant to do stuff with it... to make it yours. It has to be sized to fit. Stuff has to happen to it. It's meant to be bodged and changed after it arrives. You can see how it's been put together. It has obvious snags hanging off of it where you're meant to add stuff, or fix stuff, or attach stuff, or even move stuff around."

Bodging. What a perfect way to talk about steampunk. The word is British slang for something clumsily slapped together or temporarily patched. It implies human hands and an attitude of repair and re-use that flies in the face of a great deal of today's consumer technology. I smiled as we continued walking.

"That appeals to me because there is no technology or device that is exactly perfect for every individual." Cory held the door for me as we walked onto the school grounds. "Even if there was a perfect device for you today it wouldn't be perfect for you tomorrow. Your needs will change," he pointed at me for emphasis. "So the idea of having configurability built into the hardware and software is incredibly important."

We ordered our food. Cory kept on rolling. "Here's where my activism gets engaged. We have entered an era in the technology where the business models of technology and devices are about making it illegal for you to modify your technology. This is the business model of a video game console, an iOS device (such as the iPhone or iPad), or any large number of devices that are hermetically sealed when they get to you.

"Let's think for a second about a new device you may have just bought," Cory pulled his smartphone from his pocket. He pointed at it and narrowed his eyes at me. "Think about what you can do with your new shiny device. You can't load any software on it unless you pay for a license. The business model of the company that sold you the device is first to sell you a device, then once there are enough of these devices in the market they want to sell companies the privilege of making stuff that you can buy to bolt onto your new device. The side effect of this is a world in which no devices are configurable. For sure no devices are configurable in a way that doesn't match the business model of the company that sold you the device.

"What worries me about this is that if your needs aren't widespread enough that someone can figure out a reason to market a product to you, you're out of luck." Cory shoot his head and took a long drink of water. "There are a lot of political and technical reasons why this isn't optimal as well. I think this is really troubling. It has all these bad effects that I'm particularly interested in as a technology activist." I love this about Cory. His passion and commitment are a true inspiration. He was literally raised to save the world. His parents were activists and got their son involved with Greenpeace and nuclear disarmament at a young age. He's mainly known for his unwavering evangelism to liberalize copyright laws and his work on digital rights management, file sharing, and "post-scarcity" economics. But over the years I've come to know Cory as a humanist for the information age and beyond. He never forgets that all these new technologies and advances have an impact on our daily lives and human rights. He's not anti-technology, he's pro-human with a healthy dose of realism.

Which brings us right back to steampunk.

"Steampunk is the opposite of that," he leaned in and set his phone on the table. "It says that technology can be an artisanal, non-industrial practice, where the technology is made on human scale. It says that technology is meant to be made and modified by humans. I think this is incredibly important.

"One thing I find really troubling about the current information age is the inhumanity in our technology. It's built and shipped on such a massive scale that has no humanity in it." Cory leaned back to give room for the waiter to put his salad on the table. "We as consumers are becoming a part of a machine. This isn't new. It's the theme of a lot of industrial art and literature and movies like, Charlie Chaplin's *Modern Times*. (More on Modern Times in Chapter 13.)

"When you take humanity out of the technology, society starts to talk in a passive voice. We say things like, *mistakes were made*. We have removed people from the process so much that there is no human agency responsible. There is no human that is responsible. That's really frustrating, and ultimately really scary."

"With steampunk technology there's always a named maker associated with the object—the person who's responsible for it. You can interact with the person who made the thing that you now own. But you can do more than that. You can question their decisions or even override their decisions. Because it's an actual person you can collaborate with them to improve the technology."

Yet another important lesson. Steampunk is the antithesis of "mistakes were made," which even the *New York Times* acknowledges as a rhetorical attempt to sidestep responsibility. Steampunk seems to embrace responsibility. If a mistake was made, you know the person who made it... and it's probably you. If so, congratulations, you've learned something. This idea is a part of steampunk's inherent humanism, and something we'll continue to explore throughout this book.

The salads go away and the mains arrive. Cory continues: "Steampunk is really about a narrative of configurability. It's about changeability and mutability in technology. It embraces user modification. I really think this is what Don Norman talks about in his book *Emotional Design*.

"Norman has this idea that inherent in the nature of a complex system is that it's always a little broken. Complex systems only work to the extent that we are always fixing them. This means that you have to have a problem-solving mindset to actually get anything to work. This problem-solving mindset is improved by a good mood. No one's ever solved a problem well by getting frustrated. You solve problems by relaxing, by seeing all the possibilities and being happy. So technology that's lovely, and loving, and lovingly made, and whimsical is technology that puts you in a good mood when it breaks. And since it's broken all the time, it needs to put you in a good mood. Steampunk technology has a sense of humor. It purposely ships broken so that you can remake it for what you need as an individual."

Yes—this is scientific fact. A smile and a chuckle can make the difference between solving that thorny crossword puzzle and just tossing the Sunday *Times* into the recycle bin in frustration. Humor oils humanity's gears. And boy do we like to move. (We'll do a deeper dive into the land of laughter in Chapter 13.)

"We need to change the aesthetic of computing to make it more forgiving," Cory paused to chew and pointed at his phone. "It can't be so streamlined and hermetically sealed. We have to acknowledge that there's a certain amount of tinkering that we will have to do to get technology to do everything that we want it to do. But we have to realize that everything we might want a technology to do might include some things that no one else will want it to do. The only way you can have a technology or a device that arrives, out of the box, ready to do everything you might want it to do, is to have a device that arrives, out of the box, ready to do everything that everyone wants it to do. That's impossible. You would have to have so many capabilities designed into the device that it would be impossible to build. It's impossible to incorporate all the different possibilities at one time.

"This is really a matter of curation versus openness. Curation produces this mythical device that is shipped to you perfect. It can do everything you want to to do. Openness would produce a device that shipped to you and allows you to change it, not just once but whenever you want. I think we are seeing the ideal balance of curation and openness in the first movements towards Ubuntu (an open-source operating system; its name, which comes from southern African philosophy, means "humanity towards others.") and Android (Google's operating system for consumer devices like smartphones and tablets).

"This is a very different approach. The people who are designing these technologies still envision a device arriving to you polished to a high gleam by a team of freaky aesthetic design geniuses. It's as perfect as a pearl when it arrives. But then there's also a check box that says, 'Let me install stuff that you've never thought of'. Now that you've checked that box you might want to change something or do something different with you device. You go to your search engine and ask something simple like: 'How do I do x with y?'."

"So they're not assuming that they already know everything I might want to do with my device?" I asked.

"Exactly," Cory responded. "That search never tells you exactly how to do x with y but it does tell you something very close to x with y. It also gives you a place to ask how to customize this solution, so that you can get it exactly how you individually want it to be. Ideally, if you ask your question in such a way that it's logged, along with the answer, then the new solution can become a part of a collective knowledge that the next person who buys the same object can tap into and learn from.

"These designers are really taking the steampunk mindset and baking it into how they design technology. It's more personal and flexible. We need more of that. I'm passionate about it and I think we need to demand it from our technology."

> *Why has steampunk caught fire now, in the midst of the networked society, ten years into the relentless WTO era of global trade? I think the answer is in the motto of the excellent* SteamPunk Magazine*: "Love the machine, hate the factory."*
>
> *The factory might have given us the millionfold productivity increases that yielded the Industrial Revolution, but it achieved those gains by chaining us to machines, deskilling the artisan and turning him into a cog in the factory, stripped of judgement and dignity and disconnected to the rhythms of his spirit and the world around him. An artisan carpenter might have gone outside to sand lumber on a fine spring day when the air was blossom scented and stayed inside to varnish on a freezing day when the stove's heat defied the elements. But a factory worker can't choose her tasks or their timing; that is dictated by all the intricate interdependencies with which she is enmeshed. The factory might produce the same door at one-thousandths of the price, but at a brutal and incalculably high human cost.*
>
> **— CORY DOCTOROW**
> Introduction to *The Difference Engine*

This is an idea Cory has expressed brilliantly time and again, tapping into steampunk ideology to challenge our ideas of how we make and use the technologies around us. Let's just say that if I wasn't convinced before, Cory's made me a believer—steampunk is about a lot more than just top hats and goggles.

Remembering Just How Miraculous Technology Can Be

Lunch went by in a flash and we never stopped talking. As we walked back to Cory's office, kids were getting out of school and the streets were getting more crowded. The office was warm with afternoon light and we settled in with the clacking sounds of the passing trains behind us. In this great moment for reflection, I wondered: if steampunk is re-imaging the past then isn't the entire subculture trying to make a very different future? Then I wondered it aloud and less rhetorically. Cory?

"Steampunk tells us that our great collective unconscious is yearning for technology and institutions that feel much more human," Cory answered settling into his chair. "We want technology to be more named. We want to feel like there's a person on the other side, making decisions that we can question, or override, or at least be in dialogue with. Steampunk is really a culture about artisanship on a human scale. Not just artisanship of the material culture, but partisanship in our institutions."

This is beginning to sound a lot like an ideology, and this conversation keeps getting deeper and deeper. I'm more excited about this project by the minute. Steampunk is clearly more than the sum of its parts, yet each of those parts seems to have something to say about the future of technology.

"Steampunk makes us remember just how miraculous technology can be," Cory continued. "It reminds us that this is part of humanity's work. We are the ones who are making these technologies. They are amazing but we don't see that most of the time. We just see it as a device or technology. But steampunk, because its designs are steam and gears—technology on a human scale, makes us think about what it would mean to build a mechanical computer that could do what our computers today could do. And it's amazing. It's incredibly optimistic."

"Optimistic?" I asked. That can be a powerful word.

"Yes," Cory replied. "Steampunk is optimistic in that it mostly portrays those warm relationships the technology generates. People have accused the subculture of being a Pollyanna, because it doesn't show you the 'dark satanic mills' lurking underneath it.

"These could be the historical mills of the Victorian era, where people were slaving away," Cory explained. "Or the contemporary satanic mills of the Pacific Rim. It's important to point out that these contemporary satanic mills are where the devices and kit of the steampunk, Makers and Hackers are shipped from. I'm critical of steampunk's lack of reflexivity about where and how these objects are made. The reason that we're able to be steampunks isn't just because we glory in being makers, but it's also because of the absence of autonomy of the people who make the underlying stuff. It's a crucial piece that's missing from steampunk as a critique, not just of the material culture world that we live in, but also in that material culture underbelly that obtains in the new satanic mills. And we don't really talk about it at all." It's a valid critique, which we'll explore deeper in Chapter 9.

"We don't really talk about the blood in our technology." Cory shook his head and held up his hands. "That's a problem. We need to have that conversation. We talk about the environmental costs, the embodied energy and the carbon footprint of the stuff that we use. Steampunk certainly tries to mitigate that by recycling rather than discarding, but at the same time, there's a lack of reflexivity about the labor. We need to talk about it more."

Lewis Hine, "Mill Children in Macon, Georgia" (1909)

I wondered what it says about steampunk that Cory embraces it so deeply despite what he clearly believes to be a deeply profound flaw. More questions. Good conversations tend to spawn them.

"Ultimately for me steampunk questions whether or not we will have enough tinkerable stuff in our world. Being able to tinker and make things for yourself is very optimistic. But there's also the flip side. If someone tries to take this away we need to say, *Wait a second!* Inherently, technology should be tinkerable. When you take that away, you're confiscating something that's of value to us. Steampunk keeps alive the idea that tinkerability should be inherent in technology. That means we remain in control, not the technology, not corporations but people."

Cory wrote about this in his introduction to the 20th anniversary edition of *The Difference Engine*:

> *At its root, steampunk venerates the artisan, celebrates an abundance of technology, and still damns the factory that destroyed the former's livelihood to create the latter. Contemporary steampunk subculture inhabits a contrafactual world in which these contradictions are resolved—where "the street finds its own use for things" and where makers produce wonderments that simultaneously embody light-speed technological change and enduring artisanship.*
>
> — **CORY DOCTOROW**
>
> Introduction to *The Difference Engine* (2010)

Reflections in a Taxicab Window

By the time we were finished talking London was dark. But not satanic—the mills long-since moved to the other side of the globe. We walked over to Cory's usual bar and had a few drinks. We sat outside and watched the place fill with people as the air grew cold. At the end of the night he walked me to catch one of London's iconic black taxis and we said goodbye.

On the way back to my hotel I stared out the window. London was alive and bustling with business people and tourists. Cory had really confirmed my suspicions that steampunk really was about more than just guys in top hats and goggles and girls in bustle dresses. What's more, steampunk, along with hackers and the maker culture, had deep roots and powerful ideas. Viewed in the right light—certainly in the light of my day with Cory—you could see how it might well be telling us something important. Or at least asking new questions.

Does steampunk really ship broken? Is that the big idea we should take away from this exploration? Are we searching for a different meaning from our technology? Do we want something different from it? If so, is that something fundamentally different than what we have today? Has technology progressed to the point that our expectations have not only evolved but shifted? Are we demanding something different from our future?

One thing was clear: for me and James, steampunk had become a way to explore people's changing relationship with technology. Sure it looked cool, but there was more there. Good questions that call out to us. Steampunk really seems to be using its crazy fictions to come to a new understanding of what people want and expect from these amazing devices that we carry around with us. The past as well as the future in our pockets. Talking with Cory showed me that this active exploration of our dreams and desires from technology goes even deeper.

It's wonderful to think that the splendor and pageantry of steampunk has such serious social, cultural, and political roots that run so deep. Cultural change expressed through goggles and tops hats. Awesome.

A World-Destroying Death Ray Should Look Like a World-Destroying Death Ray

While Brian is in London, James begins his journey into steampunk culture. To start, he wants to find an expert to show him the right people to interview and the places to explore. He finds that in Cherie Preist, author of the iconic American steampunk novel *Boneshaker*.

But how James meets Cherie is a story unto itself. It's a small world and fascinating people come from all walks of life. Little did James know Cherie lived within walking distance of his home in Seattle, and they shared an unlikely mutual friend.

Three Degrees of (Hairy) Separation

James H. Carrott (Seattle, WA)

My hair is a giant pain in the ass. I put myself at some risk of bodily harm when I complain about this, but it's just too damn thick and grows out of control. For many years, while I was immersed in the world of 1960s counterculture, I let it grow long and just pulled it back into a ponytail; simple low-maintenance solution. In 2002, overwhelmed by dissertation angst, I shaved it all off. Brilliant; even lower maintenance.

Not long after I returned to Seattle from Burning Man, the women in my life finally succeeded in dragging me toward coiffure moderation. Apparently, I was told, the right haircut would leave me with a middle ground that still didn't require time and attention (my bottom line—I have better things to think about than my hair). It didn't hurt that a steampunk-themed hair salon just opened in my neighborhood. An artist friend who was active in the local steampunk scene had been to their opening gala and encouraged me to give it a shot. Urged on by the promise of a Victorian feel, clockwork knick-knacks, and a professed specialty in facial hair, I thought, "Okay, what the hell." After a few weeks of growth (yeah, that's all it took), I went in and met "Hair Guy Lancer," who's tended my noggin jungle ever since.

Lancer's active in the local goth scene, and pays attention to steampunk. He's got a natural eye for style and looks to the fringe for inspiration. Sporting a funky pompadour, a long, wispy chin beard, and a new snarky t-shirt each day—all while regularly quitting and un-quitting cigarettes—Lancer's just the right sort of odd fish to shoot the shit about just about anything off-beat and interesting. In short, my kind of people.

I went in on a soggy November afternoon (ain't much in the way of crisp Fall in Seattle) for my long-overdue cut.

"So, it's been a while, huh?" Lancer asked.

"Er, yeah," I admitted, "I suppose that's pretty obvious."

"Well, you are doing that Beatles thing... or maybe Scooby Doo. Shaggy. Whatever." He shrugged while assessing my untamed mane.

"It happens." I agreed.

"Oh, *I know*." Lancer chuckled. "Believe me..."

I shrugged a bit and said, "Yeah, well, I've been a little preoccupied with this new project I'm working on."

"Right, right..." (Lancer pulls out full-combat shears and sets to "jungle") "... another one of those cool Xbox things you can't say anything about?"

"No, it's actually about steampunk." I said. "I'm studying it with a friend from Intel."

"What?" (Stops cutting to give me an incredulous look) "Stop shitting me."

I was relieved he'd stopped cutting when he said that. "I'm totally serious. You know I'm a historian, right?"

(Clippers back to combat mode) "No. Wait, how does that work with the whole, like, high tech thing?"

"I kinda defected," I said sheepishly. "Long story. But you can see how it fits with the whole steampunk thing."

"Mmm-hmm," he hummed. "Totally."

"Well, my friend is a futurist," I continued, "and we're doing some research on what steampunk can teach us about the future. I didn't see you at Steamcon, by the way. Were you there?"

"Dude, I'm in school *and* I work all the time," Lancer replied. "My god, I barely even sleep."

"I hear ya. I missed it last year 'cause of work, but this year I had to go." I looked around the salon. "You know, it would be awesome to talk to you about this place sometime. How you guys came up with the theme and so forth."

"Well..." (Pauses cutting to think) "I could introduce you to our owner, who did all that. But, you know who you should really talk to?"

This sounded like it was gonna be good. "Who?"

(Stopping the clippers again...) "Dude, sit still or you're gonna lose an ear or something."

"Sorry. Sitting still's never been my strong suit."

"Oh, *I know*..." Lancer chuckled.

"But *you* made me curious. Who should I talk to?"

"Well, Cherie," he said. "Cherie Priest."

"Oh, totally!" I jumped. "I know she lives here in Seattle, but I haven't had time to go through the whole 'talk to the agent' thing yet."

(Ouch)

"Dude... sitting still, remember? It's cool, it's cool. No blood or anything but you gotta chill out. I do her hair too, you know. She's awesome. I can introduce you on Facebook. She's a friend of a friend from back east."

Okay, fine, maybe I really did need to chill out. But I had good reason to be excited—Cherie Priest was absolutely someone I needed to talk to. And now maybe I could!

By this time I was into steampunk literature in earnest, and had just finished reading Cherie's first steampunk book, *Boneshaker*. The book had really drawn me in with its near-apocalyptic vision of Seattle, and I couldn't get its imagery of a struggling underground version of the city, fraught with the practicalities of containing deadly gasses and a zombie-infested surface out of my head. The way she wrote about technology captivated me, too—the artifacts and gadgets in her world were almost characters in their own right.

If I hadn't yet been convinced that the world was truly a thousand times smaller than I had once thought, here was further evidence. Lancer did Cherie's hair, too; a brilliant trademark blue that graced many a book jacket. Lancer introduced us as promised, and Cherie turned out to be one of the kindest, most accessible, and just plain excellent human beings I've ever encountered. I explained the project and we made plans for coffee in early January, once our respective holiday hullabaloos had settled down.

I met Cherie for coffee at Victrola—a short walk from home for each of us and among the better coffee shops in Capitol Hill. It's a bit hipster-y and can get crowded, but they know that a good mocha doesn't need to be enamel-melting sweet and that goes a long way for me. We found seats on a couch at the back, next to an old piano which I would have been shocked to hear had been played in the past decade. I had prepared a long list of questions, nervously going over and over again in my head how I'd draw out the brilliant insights she was sure to have.

Well, no worries on the brilliant insights front, but turns out my nerves were entirely unwarranted. Cherie is nothing short of completely charming. We spent a couple hours talking about the American Civil War, Confederate patent archives, re-enactors, alternative history, and the myth of the South. In short, fantastic stuff, but we didn't even have time to turn to steampunk. While we live within an easy walk from one another, it took a steampunk convention in Portland to finally get us together for a formal interview.

From Elder Goth to American Steampunk Icon

Cherie gets it. I mean really gets it. She's been on the steampunk scene for years now, and not only understands the subculture itself, but is aware of her own role in it.

A Right and Proper Interview: James with Cherie Priest at The Old Church, Portland, OR July 2011 (photo courtesy of Ben Z Mund)

Since its publication in 2009, Cherie Priest's novel *Boneshaker* has become the popular fiction poster-child for contemporary steampunk. Its iconic cover appears in book shop facings, and anchors display shelves at libraries across the country. With both popular success and subculture cred, her "Clockwork Century" books are becoming part of the new canon of American steampunk.

Since this was to be a filmed and photographed interview, the convention hotel was hardly an adequate venue. So we met up at The Old Church; built in 1883, the aptly-named oldest church in Portland. Hemmed in by stained glass and paneled wood, and pinned like deer by the lights and cameras of Byrd McDonald's documentary crew (yep, we'd become a film as well—just too much good stuff for a single medium to contain; more on this in Chapter 5), we switched our iPhones into silent mode and set down to the business of discussing yesterday's future.

"Let's begin at your steampunk beginning," I said. "When did you first hear the word *steampunk*"?

"About four or five years ago," she replied without hesitation. Again with the mid 2000s...

"I slid into steampunk sideways as an elder goth," Cherie continued. "There's a lot of overlap. I was poking around on the internet and I came across a forum, which will remain unnamed. There were a number of teenagers who were complaining about Americans pretending to be steampunks. They were getting all riled up because, obviously, you can't have steampunk set in America. Steampunk has to be about Victorian England. And I got thinking about it, reading their list of reasons."

"Well we had colonialism, check," Cherie continued. "We had all this technological revolution, this industrial uprising, check. We had wars, check. We had the socioeconomic disparity, check. I came very close to actually posting on this forum to say, 'We liberated an entire slave class with no social or legal protection and turned them loose in the middle of a Civil War. I don't know, I think maybe we have the social disparity and injustice covered.' And that's not something to be proud of, but certainly true. So, when people said that steampunk had to be only about Victorian England, I thought they were being ridiculous.

"It made me so mad I wanted to prove that there was no good reason you couldn't have American steampunk. And don't get me wrong, Victorian England is a wonderful place to set steampunk. It's just not the *only* place. So that was probably maybe 2007, 2008. And I started working on *Boneshaker* later that year."

Clearly Cherie's a path-blazer too. We'll dig deeper into what I've come to think of as "The Victorian Problem" in Chapter 8. For the moment, it suffices to say that this whole Victorian Britain thing is one heck of a sticky wicket—good thing Cherie's a bloody brilliant batswoman. Er... cricket metaphor anyone?

"So," I returned to the point, "it's 2008 and you're writing American steampunk. How did you explain that idea to people? Say you bumped into someone you hadn't seen since high school—what would you say about what you were writing?"

"Oh, I had to explain it to my dad in 50 words or less," she replied quickly. "Now, I have it down. Steampunk is a style. Books, music, video games, movies, what have you—that draws inspiration from the science fiction of the 19th century, like Mary Shelley, H. G. Wells, Jules Verne, those guys. It is retro-futurism. It is the science fiction of the future that never happened.

"That usually does it," she smiled.

I allowed as how it darn well might. But I needed to push a little more, because she'd raised a pont that had really snagged my interest. Steampunk really does draw a great deal of its inspiration from "The Age of Steam." Why? There's a whole lot of history, technology, and imagination both before and after steampunk's 19th century—roughly 1818-1918.

So why then? And why then, *now*? The latter is admittedly an awkward question, but past-punking can throw a curve ball into one's grammar (metaphor shift from cricket to baseball... y'all Americans more comfy now?). It ain't simple referring to the idea of a time period referring to a(nother) time period. I decided just to shoot simple and straight.

"Retro-futurism can be a lot of things, though, right?" I asked. "Why this latching onto the 19th century in particular?"

Cherie paused to think for a moment. She's not only a steampunk thinker *par excellence*, she's a rare blend of open, thoughtful, and brilliant—the kind of person who's not afraid to think out loud and whose thought process teaches you something with every step.

True to form, her response hit an important nail on the head: "I think one reason that we keep going back to the Victorians, and why steampunk has glommed on to them, is because there was such enthusiasm and such optimism for technology. They honestly believed that the technology was going to save them. Mary Shelley thought maybe it could raise the dead."

Yep, this steampunk thing has *everything* to do with technology.

"So there was this idea that we're on the cusp of greatness," Cherie continued. "We're on the cusp of saving humanity and conquering the universe and doing exciting things, whereas the science fiction of the last 20 or 30 years has been more to the tune of 'we're going to destroy ourselves with our technology.' With cyberpunk, and even a lot of the space opera from the '70s, you get this idea that technology is not here to help us—it's going to be the end of us. And I think we look back on the Victorians now and we see a very refreshing attitude, like, *Isn't technology exciting? And isn't it good?* Whether it is or not they certainly thought it was."

A breath of fresh air in a stuffy cyberpunk room? A new way of thinking about our machines and ourselves? Those are really compelling reasons why. And "Why?" is nearly always the most important question. Next to how, it's the keystone of historical inquiry—who, what, where, and even when are window dressing in comparison. Cherie just handed us a really interesting why.

"So, what is your relationship with technology?" I took the opening to tie the idea back to her life now. "Did you ever think that one day you might be perceived as a bit of a technologist?"

"Honestly," Cherie admitted, "I've always been a little bit of a technophobe in some respects, but I like video games. The Atari 2600 was my first system. I loved it and I played it to death. Math and science were not my thing at all, but I find them terribly interesting now. I have a ludicrous fangirl crush on Neil deGrasse Tyson from *NOVA scienceNOW*. He's my hero, because he doesn't dumb it down—makes it interesting, but it's accessible. And I don't have to feel like a moron for trying to delve too deeply into something that I don't understand."

I sure can't argue with nerd crushes and I'm all about bringing the highfalutin down to earth. Got that. But here we are surrounded by intricate stained glass and gorgeous carved wood and we're talking about video games. Makes sense to me, but how does Cherie connect this with her steampunk world?

"So," I asked, "what is it then? What drives all the really cool, detailed gadgetry in your books? I mean, Boneshaker and Dreadnought certainly don't read like they're coming from someone who says she's *a bit of a technophobe*."

I, of all people, should have anticipated her answer, but I didn't. It was a problem of Yankee perspective.

What If They Didn't Run Out of War?

"Well, I grew up in the Southeast where, let's be honest, alternate theories of the Civil War are kind of a regional pastime." She delivered this with a knowing smile and a wave of her hand.

Yeah, just a wee bit of understatement there.

When my Union infantry unit went south to participate in a 125th anniversary re-enactment, we participated in a parade through the small town of Calhoun, Georgia. Some 3,000 Confederate troops marched through the town to flag waving (red, white, and blue—but the stripes weren't horizontal) cheers and adulation. When our maybe 1,000 men in blue rounded the corner onto the main street there was dead silence. All we heard as we marched through town was the sound of our own steel heel plates clacking on the pavement. Let's just say that General William T. Sherman didn't leave a whole lot of friends behind in Georgia.

Cherie knows this well. But she also knows that the Civil War left behind a lot more than scars. "I went to the University of Tennessee for my graduate school working in Chattanooga," she continued. "You can go through the old patent archives there, and find crazy, crazy patent submissions for machines that people wanted to make. These crazy war machines..." she mused, clearly still inspired. "Would they have worked, who knows? Probably not. But they were very interesting. And they were the most steampunk things I had ever seen and I thought, I would love to use these. Except they never happened, because they ran out of war."

"So it started there. Well, what if they didn't run out of war? What if that kept going? Because nothing drives technology like war."

Too true. In fact, we owe a great deal of our contemporary information technology to the innovations of Allied code makers and breakers in places like Bletchley Park and Bell Labs. But I'm jumping ahead again. Time travel. It happens.

Cherie continued, laying out an elegantly simple and deeply brilliant premise: "I thought, 'well we'll give them another ten years to start developing these things and to start making these crazy machines and get them off the ground and go from there.'

"But that meant that *Boneshaker* in particular was easy in a way, because it's set in the Pacific Northwest. And let's be honest, the Pacific Northwest didn't see a lot of Civil War action. The number one piece of feedback—disguised as complaint —" I suppressed a little laugh, Cherie's got a flair for turning criticism into compliment; a certain Southern conversational jiu-jitsu "—that I got from *Boneshaker* was that people wanted to see what the rest of America looked like. So when it came time for *Dreadnought*, I thought: *well okay, let's do that.* So I got to start in on questions like: *How did we get a 20 year Civil War? What would have needed to happen?*

"Yes, the technology is what drove the books." She punctuated, seeming to think we'd digressed a bit. Perhaps we had, but it was all a part of the whole for us both. "But do I pretend to have any serious intense, advanced technical knowledge and understanding? No, I don't."

Ah, but we're talking steampunk here. And, sharp lady that she is, that's precisely where Cherie went.

"When it came to steampunk, it just didn't seem to have a mythology to account for all of the minor tropes of it. It was a collection of things that people wore, or a collection of things that people watched or listened to. Like goggles, a gas mask, was kind of a big one. There's so much that's borrowed from fetish culture as well as Goth culture and the maker movement and the rest. And I wanted to give it a unifying mythology. Which was tricky, but I thought: *Okay, gas mask. Well, we need gas. Well, what kind of gas? Well how do these things work?* And I tried to make technology that was symptomatic of the place, because I didn't want it to be gears on hats. I wanted these people to have things that were going to be practical to them in a very strange environment. And if it was completely ludicrous, then, well you know, it's fiction."

It's this degree of thought that makes Cherie's work so compelling. Zombie gas in the walled-off streets of a long-lost Seattle? Sounds nuts at first blush, but dig in to *Boneshaker* and you'll be surprised at how well it all comes together. Good steampunk does that—technology, story, setting, and characters entwine in a way that creates a whole greater than the sum of its parts. You may have to take a big logical leap back and to the side to get there, but when your feet hit that new ground, it'll open your mind like few other things can. Unlike hard science fiction where the technical details of a given device or craft are meticulously researched and presented, steampunk's tech tends to function at times on quirkiness alone. You get what you need to fuel the story... and your imagination. The last bit is really the point.

"So with regards to the technology," She continued. As my mind wandered into the fantastical streets of my home gone historical rogue, Cherie—who paved them—stayed right on track. "I wanted it to be symptomatic of the setting and of the needs of the people who live there. And once I figured out what their technology needed to do, then I was able to, with some hand waving, reverse engineer what that would look like or how that would come into play."

Hand waving indeed. Stage magic extraordinaire.

"I keep adding these disclaimers to the subsequent Clockwork Century books, because I get so much hate mail from people complaining about how badly I've mangled history. And it never ceases to amaze and amuse me. Especially when I get the ones from people who are complaining that I've used a building that wasn't built until 1898, or something. I'm like, 'really, you just went right past the zombies. And you were OK with that? But you got hung up on this one detail.' It's for fun, and I feel like *if you're not having fun, you're doing it wrong.*"

I added the italics (people, even those with Cherie's rare aplomb, don't speak in italics), but the emphasis is hers. It's a message I've heard her deliver time and again at conventions and readings, and is also one of the few places where I've become a vocal partisan. As we continue this adventure, we'll see lots of definitions and assertions. Everyone has their own pet notion of what steampunk is, but as my friend Howard Zinn used to say: "you can't be neutral on a moving train" (heck, it's the title of his 1994 autobiography). What Howard taught about traditional history applies even more strongly to a participant-observer approach. When you get involved in a community, you get involved in its struggles and arguments. Silence just puts you in implicit agreement with whoever's shouting the loudest.

Steampunk is absolutely a moving train and I won't pretend to be neutral. Cherie is right. No matter what issues your steampunk digs into, no matter how passionate you feel about any element of your imaginary historical expedition, *if you're not having fun, you're doing it wrong*. Well over a year later, as I edit this interview for publication, I still can't come up with another reason I'd ever tell anyone that they're doing steampunk wrong. Maybe when it attacks another person's steampunk, but that's sure no fun for the attacked.

I digress. We'll get back into these issues later on. For now, let's just say that Cherie lit a fuse in my heart here.

I went back to the central reason I was here in the first place. Steampunk was taking off, and Cherie had a view from the cockpit. "There's something resonating about what you're doing in the Clockwork Century books. They're taking off, and people are clearly responding to them. What do you think," I asked, "about how the ideas about history and technology you've been talking about are echoing? How people are receiving them? What is it that people responding to? Is it the technology?"

"I hope it's because they are stories about people, first and foremost." She replied. "The first wave in any new genre or any new subculture—let's use gaming as an example—are always the people who want to show you their character sheet. It's: 'look at this world that I made, look at the stuff that is in it, look at the gadgets that I have, look at the toys that are in this world.' And that's great, but I'm not so much interested in that, I would like to know who lives there.

"*Boneshaker* is kind of a claustrophobic book, and it's intended to be. It's only a handful of characters over I guess about three days. And it needed to be a very personal story. Because if it's not a personal story, then it's just not the kind of thing that I plug into very well. And so even when my technology is grossly, grossly awry, at least I hope that the people ring true.

"I feel like I would be remiss if I didn't name check Charlie Stross at this point, because if I ever meet him I'm going to just buy him drinks and send him flowers. He did one of those, 'hey you kids, get off my lawn all you steampunks' posts and he called me out by name and linked my website. He sent me more traffic than I had gotten in years by saying something to the effect of how steampunk doesn't really seem to have any actual credible technology in it. And then in parentheses he said, 'Cherie Priest I'm looking at you with your gas-powered zombies.' People sent it to me, you know, *oh my god, have you seen what Charlie Stross had said about your zombies?* And the only reaction I could have was: 'Oh my god! If only I had consulted more zombie scientists. We could have avoided this whole embarrassing—"

Note

Science Fiction author Charles "Charlie" Stross's blogged critique of steampunk, titled "The Hard Edge of Empire (*http://www.antipope.org/charlie/blog-static/2010/10/the-hard-edge-of-empire.html*)" kinda took the steampunk world by storm in late 2010. In it, he called out the literary genre for its dangerous tendency to romanticize empire (and was right to do so—more on that in Chapter 9). He also, as a writer of "hard" science fiction, took vocal issue with steampunk's spurious notion of "science." Much debate and kerfuffle followed in the wake of this influential post.

Being in no way a zombie scientist, I could hardly top that. "Speaking of attention," I said, "there's so much activity around steampunk right now, so many conventions popping up, so much being written, and so many cool hunters grabbing onto it. Why now? What do you think are the reasons why now is the time that steampunk is really becoming cool?"

"I don't think there's one reason," she responded thoughtfully. "I think you get the post-millennial thing, you have the rise of the maker's movement, you have the rise of environmentalism on a much wider scale than it has been before. Sometimes I will half-joke that if steampunk has a philosophy it's: 'Reduce, reuse, and recycle.' It's: 'Find new purposes for old things. Find the beauty in old things even if they don't work anymore.'

"And I think it's partly because, especially in the last 10 or 20 years—and I say this as someone who didn't have an e-mail address until 1998—as our technology has become so terribly, terribly powerful and we've become so dependent on it, there's simultaneously something very fragile about it. I mean, I have an iPhone that we call 'the Jesus phone.' I can see myself from space with that phone, but God help me, if I drop it in a bathtub. I mean it's all over."

I get this instantly. I had recently dropped my own iPhone in the sink and paid hundreds of dollars in idiot tax for that simple slip of the hand.

Cherie continued, revealing more complexity in her self-portrait (technophile and technophobe aren't quite so diametrically opposed as we assume). "When the grid goes down—and I'm inconsolable if my internet is down for ten minutes—but if the grid goes down what do we have that still works? Well, we have the old things, we have the analog things. We have the things that we don't have to charge. The things that we don't have to need wireless connectivity to use. And again maybe I think about this partly because I grew up in the southeast and around Civil War re-enactors and historians, where they have these hundred fifty year old guns that work beautifully and they use them at demonstrations. And they have these old machines and old gadgets and these things still work. They were built to last.

"Those things will be there when the zombie apocalypse comes. And I think, that's part of it, too. The rise of the zombie trend, if you will, for the last five or ten years has coincided along with the exact same thing. It's this fear that we have become too dependent on our technology and when it goes away what do we have left?"

Yep, but what's technology? Where's the delta between old gadgets and new? Is this really about fears of computers and the internet? Rhetorical musings for the moment, but we'll come back to these questions later in the book.

Like any responsible adult, I am concerned about the impending zombie apocalypse. But I'm also a historian. "So," I asked, "how does that play with that Victorian sense of optimism about the future? From a historical perspective, I see some irony in that, and I know you get this too, because what was happening in the Industrial Revolution? Well, there was a class of people who were very excited about all this progress and opportunity. There was also a class of people who were being very oppressed by all of that at the same time. So how does looking backward and applying a kind of Victorian optimism to the future work? How does it all play out for you?"

"Well I think that's what separates us from them." She drew a useful and interesting line. "When you're dealing with the Victorians and their optimism and excitement about the future, they didn't mean for everybody, they meant for *us*. For us right now and we're not really paying attention to all those little people over there doing those little things.

"This is something I've found myself defending a few times when I'm confronted about idealizing the Victorian era: 'well don't you know how bad it was? How horrible it was for so many people?' Yes, and we're not trying to return to the Victorian Era. I like deodorant and soap and I like Wi-Fi. I don't want to go back, but there were good things that can be salvaged from that time. There were good ideas and there were good people and there were good sentiments and there were good philosophies.

"You can't throw out the baby with the bath water, with regards to anything. And again it comes around to 'reduce, reuse, and recycle' and 'find the value in something that has been discarded.' And so we're kind of combing the Victorians to see the things that we like and the things that still work."

There's Room for Everyone to Play

It seemed a two-way street to me. "Are you putting stuff back into history to change that too?" I asked. "I'm thinking about characters like *Clementine*'s Croggon Hainey, a former slave who captures a Confederate airship and becomes a renowned pirate, and *Boneshaker*'s Briar Wilkes, who's right up there with Ellen Ripley—an undeniably strong, heroic mother figure."

"Hainey was fun, and yes, Briar's a total homage to Ellen Ripley and Sarah Connor."

"Is there part of writing these characters that feels political? Or is it just that you're a 21st century woman who wants to write about strong, contemporary characters?"

"Well now I'll give you this anecdote," Cherie replied in pure southern style, "because it sums everything up in a nutshell. I moved around a lot when I was a kid and I spent a number of years in Texas. And my running joke with Texas, which I love by the way and still have family there, is: 'You know how cats used to be worshipped as gods and they've never let you forget it? Texas used to be a republic and it's never going to let you forget it.' You have to take Texas history every single year that you were in elementary school. So one year I took it and I had read the little book. It was a multi-grade classroom I was in, so I was reading one of the older kid's books. I got to the end of it and I thought: 'that is so cool. All these cowboys and all these crazy things going on in these wars and rawr this looks like fun.'

"I was maybe ten years old, but something occurred to me and I went up and I asked my teacher, who was not a Texan, I would like to be clear. He was a mean old man from New England. And I said: 'how come there were no ladies in Texas?' Because it had dawned on me at the end that there hadn't been any women anywhere in the book. And that mean old man just does this 'Ohohoho, there were ladies in Texas; they just weren't doing anything important.' And I'm ten. It's not like I have these really finely tuned feminist sensibilities, but I took that a little personally, because that doesn't just mean that women back then weren't doing anything important, it means that women now probably aren't doing anything important and you're not going to do anything important."

She did the voices too. It'd've been precious if it wasn't so darn true. Women have been written out of a lot of Western (the kind followed by "hemisphere," not preceded by "Wild") history. That's changing as new generations of historians and teachers take the place of folks like Cherie's mean old man from New England (here *I* would like to be clear that I know a great number of very nice folks from New England), but the "great battles, great patriots, great men" school still holds a heck of a lot of sway.

Even at ten, Cherie was no fool: "But obviously there were lots of women doing lots of important things. And furthermore there were lots of other different kinds of people as well. There were a lot of people of color. My god, the immigrant situation, with people coming from so many places to fill up this vast expanse. And the Native American population. All these different kinds of people, who you just don't see, in the history books as we tend to understand them.

"There was kind of a sense to me," she continued, easing back into her steampunk craft, "in reclaiming that. To go: 'You know what? We were here and we were doing things. And those things were not just important, but maybe most important of all, they were things worth playing with. And we have a right to be here, we have a right to be part of history.' And it doesn't matter who you are. I've been accused of you know hating on white guys and I have to say: "No, no, no, it's not even just 'white guys'." God help you if you were Jewish, if you were Catholic, if you were Irish, or Italian. Or if you didn't speak English."

Stop and think about it for a second and you may get what she means. What the heck is "a white guy"? Come to think of it, what's "a black guy"? These are lump sum terms that *seem* to make sense until you look at them closely. A book I read in grad school jumps to mind: Noel Ignatiev's *How the Irish Became White*. Title says it all. Steampunk gives us another tool to dig into the past and ask these kinds of questions—to highlight the strange assumptions we make about who we are through reaching back and challenging our ideas about who we *were*.

Part of the reason I enjoy talking to Cherie so much is that she's a darn savvy historian in her own right. "Depending on where you came from," she said, "the dominant story had a very, very, very narrow definition of who was valid. I reject that reality and I replace it with the one that was actually there. So when I do these alternate history books, I want to make it clear: these are alternate history, they're not an alternate dimension. So we have lots of different kinds of people here cause here in the real world we have lots of different kinds of people."

In other words, her history is made up, but it is in some ways more true than what you'll read in a textbook. Cherie's not alone in this. Anina Bennett and Paul

Guinan's *Boilerplate* is spectacular. We'll talk with them soon. Let's get back to how Cherie does this, though, because it's a fantastic example of how you go about messing with the past the right way—having fun and thinking at the same time. Doing your homework and then playing with it.

"Is it a little political for me?" she continued. "Yes, but is it completely political? No, because the characters are the interesting parts. And I don't want to limit myself to just the expected kind of character. I want there to be room for everyone to come and play in my alternate history universe."

My heretical historian's heart did a little leap of joy. Playing with history. Even more—history that's "worth playing with." Gold, Cherie. Thank you.

I switched back from internal monologue to external dialogue: "We've been talking about some important stuff, but at the same time, you keep talking about play. Is a lot of this about play? Steampunk and messing with the past?"

"Well, sure." She responded to what was at this point in the conversation a slow fly ball. "As I've said, if you're not having fun you're doing it wrong. People who try and take steampunk really, really seriously, who want to be the arbiters of taste, who want to be the people who define the parameters into which this has to fit—honestly, I don't have a lot of patience for that."

"So are we taking ourselves too seriously?" Was I asking about steampunks or our culture in general? At this point "we" was starting to blur for me. Fortunately, Cherie answered for both my Jekyll and Hyde.

"I hope not. Most of the steampunk events I do, it looks like everybody's pretty much there just to have a ridiculously good time. Obviously there's a lot of serious conversation, too: I mean a good number of academics are leaping into this. Especially interesting, from my angle, has been the people who are re-appropriating colonialism. That I think is fantastic. It's like, yes, look you were all here, too. Come and play with us. There's room for everyone here."

"Is there something particular about steampunk, when it comes to this idea of being inclusive?"

"Well, back in my elder goth days, there was this broad perception that if you were a goth, that was the pasty white boys club. And I was usually outnumbered by boys four or five to one. But when it comes to steampunk the appeal is so much broader I mean it's about 50/50 men and women and you see people of all ages. The steampunked baby carriages just slay me.

"At the first SteamCon a couple years ago the hotel reached its capacity. The fire marshal cut them off at a certain point, and they couldn't put out any more badges. But there was this older woman whose mother was with her, who were just there staying in the hotel. Little tiny old lady easily in her 80s. She comes downstairs and is looking at all the costumes and she was very, very interested in this and since she was already in the hotel, it didn't matter and they gave her a badge and somebody ran her through the dealer's room and tarted her up. And she came out feeling fabulous and they both joined the party. I mean, there is room for everybody."

We'll find out just how true this is as our adventure continues throughout this book, but as a responsible interviewer, I couldn't just get caught up in starry-eyed possibility.

"And yet there are times," I said, "when an outsider to all this might just look at all this steampunk stuff, especially thinking about it as science fiction, and be bored to tears, right? All this potential to dream and create anything you can imagine and yet we often just end up looking at a lot of brass."

"Yeah, a lot of brown."

"So, what is it about the aesthetic you think is catching on with people?"

TECHNOLOGY SHOULD BE BEAUTIFUL... AND HANDS-ON

"I think it's quite simple, actually," she replied. "Are you familiar with the Oxygen School of Design? Think about an Apple product—it's an inscrutable brick, unless you know what to do with it. In the last ten or fifteen years in particular there has been this movement in design towards making your technology flat and inscrutable.

"The Victorians would not have bought into that for five seconds. They thought if you were going to make a giant death ray killing machine, it should fill an entire room and it should be gorgeous and it should have a million levers and buttons that don't even do anything they just look cool, but it should look like a giant death ray killing machine. And now we have a button. I think steampunk is a rejection of that. I think it's embracing the idea that your technology should be beautiful and its beauty should match its capacity for power or interest or what have you."

"What about this idea of taking things apart and tinkering with them? Is this all just the stuff of fantasy, or is it realistic that we should expect this of the objects, the technology in our lives? There are certainly people who'd say that this kind of tinker-think is just a pipe dream, given the powerful consumer culture we live in today."

"I think if that's what they think... he seems to be having a wonderful time with it. Making all kinds of interesting things. And not just pretty things, useful things. Things that actually work and things that you can actually play with."

Yeah, I know. Cory Doctorow might've kicked me in the verbal kneecaps for that question, but ya gotta push. We interviewed Jake Von Slatt as part of this project. We'll repay the karmic debt on the maker score, never fear.

"I think," Cherie continued, " that part of that appeal—the analog appeal and the hands-on tech, I want to do it myself appeal—is this idea that, well like I said, if my iPhone broke, I couldn't fix it, not in a million years, but if a giant death ray killing machine broke, I could probably fix it with a wrench and a couple of rivets. There's something very reassuring about that.

"And maybe the grid won't go down. Maybe we will end up in some vast space exploring a science fiction future and that'll be fantastic, but the hammer and nail is never going to go away and there's always going to be a need for people who are interested in that kind of thing. I think as specialization has leaned toward high tech, especially in this country, we're starting to lose a lot of that, we're starting to leave a lot of that behind. I'm not sure that's a good thing. I think it is good and useful and helpful for people to be able to maintain their own property and maintain things that they care about. People who can do—"

"Are sexy." Sorry, but it's true.

"Exactly. It's a part of the same thing I think really, because not everyone can program computers. Not everybody knows all these languages. Not everybody can land one of the last good-paying jobs that are still around here. I think in part it's a fetishizing of there being a time when you could earn a good living with your hands. And that time seems to be passing. As for me and my blue collar characters, I mean my dad was in the army and my mom was a schoolteacher. Most of the people I knew were blue collar. So those are the people I write about.

"So is there something in the fact that so many of us spend so much time in front of computer screens these days that drives this urge to actually stand up, walk across the room, and pick up a wrench?"

"Sure," Cherie replied. "You know who talks about this quite a lot? Mike Rowe who hosts *Dirty Jobs* has talked in front of Congress and has done a lot of public speaking about the return to working with your hands and blue collar work. And how it is not only important, but it's satisfying. And that maybe it scratches an itch that humanity needs to have scratched."

"Humanity?" Got me again. Totally intrigued. "How?"

"Yes, humanity. Because we've gotten away from it. Just the other day, there was some study that made it online about how sitting all day is bad for your health, even worse than smoking. And the sedentary thing that the vast majority of us who fancy ourselves professionals, we're either in a classroom, or we're sitting at a desk, or we're in front of a computer, whatever it is we're doing all day. But we don't have the satisfaction of having accomplished something physical that we can point to and say, 'I did that.' I had a friend who was a coder who used to complain about that: "I spent 14 hours coding this site and making sure every little thing worked and it was really fantastic. And I can't show anybody, unless I have Wi-Fi and a power supply.

"So there is something in the immediate gratification of having made something. People are creative species, to varying extents. People like to make things, they like to do things, they like to share things. Art is immensely powerful. They use it for school children, they use it for prisoners, they use it for post-traumatic stress survivors. It's a way for people to interact and to communicate that is easier in the very painful things. So it's important that people are able to make and produce and do things. And we have increasingly become a society where we consume things: we don't produce things. And so this is a reaction to that. This is kind of a coming around to that: 'no, but I want to make something. Even if it's only pretty, pretty is its own reward.' So let's try that and play."

Pretty Is Its Own Reward

The Old Church, where we held the interview, was just a short walk from my hotel. I swept back down the warm sunlit streets (Portland has even more July than Seattle) just as I'd come, burgundy frock coat fluttering in the wind, and thought of Cherie's death ray machine. Form and function, past and play... she'd given me a lot to think about. And it all started with a haircut. Makes sense, when you think of it—all tangled up with contradictions and snarled with questions, but as Walt Whitman (one hairy old dude himself) reminded us, so are we. Steampunk is large. It contains multitudes. Clearly, we'd just touched the tip of the iceberg, but patterns were beginning to form. Lost in thought as I strolled, I only barely noticed the looks my costume garnered from passers-by. Yep, it was time to kick this thing up a notch. I called Brian and we set the next ball in motion.

5

Steampunk: A Dinner in Three Courses

James and Brian dive headfirst into steampunk. To get a better grasp on the culture, the boys collect a group of steampunks, makers and hackers to get together and have a spirited dinner. Each course has a different topic of discussion. When the wine starts flowing, not everyone agrees, but the conversation is fascinating. To capture it all the boys call independent film producer and documentarian Byrd McDonald. He brings his crew to Seattle and at the end of the dinner they become a part of the Vintage Tomorrows family!

Ultimately they are all looking to understand: What is steampunk? What can steampunk teach us about the future?

Getting the Band Back Together

James H. Carrott and Brian David Johnson (Seattle, WA and Portland, OR)

MUCH OF THE research we do is simply talking to people. Over the course of researching this project we got to meet and talk to so many interesting people. Years ago as a part of his research Brian started getting as many of the smart people as possible together to sit down and have a dinner to talk about the project. It's amazing to get a table of passionate people together to talk about a single subject. The research almost does itself. It's a pretty amazing thing to see. It's also a whole lot of fun.

Brian was explaining this to James during one of their working sessions. James was poking around the stacks at the Seattle University Library and Brian was in his car driving home from Intel.

"So we get them all together and get them talking about what we learned," Brian shouted into the hands free microphone in the car. To be heard while Brian was driving he always had to shout. James never complained about it for the entire project. This makes him a saint, though a slightly wincing one—you're supposed to be quiet in libraries. "During the dinner we put an audio recorder in the middle of the table. When I did it with a group of video game developers it went great."

"I love it," James replied. "It'll be good to get them all together and talking. Some of them really don't agree with each other. I bet we'll get some interesting stuff."

James has a penchant for disagreement. It gets him in trouble from time to time, but he's rarely bored.

"Perfect!" Brian yelled. "I was thinking we break it up into three courses: appetizer, main course, and dessert. Each course has a different question or topic for discussion!"

"What did you have in mind?"

"Well first we start with something easy like what is steampunk or what does it mean to you. Then maybe we add something like how did you first come to hear about it and why were you attached to it."

"Easy?" James laughed, quickly checking his back for disapproving librarians. "That could take up the whole night! These folks love to talk, and nobody agrees on what steampunk is."

James was already assembling a mental list for the dinner—a really diverse group of people. He'd really spent the majority of time with these folks during the research. Though he'd been behind the scenes from the start, Brian was the new guy. Most of them had never met him.

"What do you think the second course should be about?" Brian asked.

A brief pause, then James went where you might expect a historian to go: "Why now? Why is this particular subculture important at this point in time?"

"I agree!" Brian replied, then bookended with his own expertise. "I really think the last course—the dessert—should be about what steampunk can teach us about the future. We have learned so much from this research about how our relationship with technology is changing and how people want it to change. "

"Some are even demanding that it change," James broke in. "That's what the makers are all about. Change, I mean, not dessert. Though I don't know anyone who doesn't like dessert."

Another pause, then some road noise...

"James! James! James! Wait I just thought of something. These are steampunks. They are going to dress up!"

"Well yeah, of course. They're steampunks. I'm going to wear some of my old Civil War re-enactment gear. I've modified it a bit and it works surprisingly well with the new arm garters and top hat I picked up at... "

"That's great James, but you are *all* going to dress up..." Brian is clearly working out the implications as he thinks out loud.

"Yep. What are you going to wear?" We were clearly talking past one another. It happens.

"I'm a futurist, not a steampunk." Brian is serious here.

"True. But you could..."

"James, James." Brian deployed his *get to the point* voice. "That doesn't matter. I was thinking. We're going to have a table full of steampunks. We have to *film* this!"

"That's brilliant!" James yelled back, ducking behind a tall shelf to avoid the sharp stare he'd attracted from a studious bibliophile.

James made it out of the library without reprimand, and Brian safely finished his drive home.

The prep for the dinner went surprisingly smoothly. James got everyone together and enlisted the help of his ex-wife, Margo Robb—a fantastic organizer—to find us a restaurant that would: a) let a group of Steampunks take it over for a night; b) be okay with a pretty large film crew and all their gear; and c) do all this on Easter Sunday. This round's sainthoods went to Margo and the good folks at the Dahlia Lounge.

Right after we had finished up our planning, Brian called Byrd McDonald. He'd known Byrd for years. They had co-directed a documentary together back in 2004 called *Haunters*. (That's right, our futurist is a filmmaker, too!) It was about people who put on haunted houses. Not real haunted houses. These were people who pulled together pretty massive spectacles for Halloween and charged people to go through them. People love to be scared. We've got reams of footage to prove it—folks terrified and loving it. The spectacle of the houses was incredible. It was like regional theater but with fake blood and zombies.

Since then Byrd had become a successful independent film producer. Brian knew this project would really catch his imagination. He called Byrd and told him that he had to come up to Seattle and bring a crew to film the dinner. "I didn't know what we'd do with it but we had to film it. Come on!" Brian pushed. "A table full of steampunk—what part of that is not going to be awesome?"

Byrd laughed and agreed. He'd be there. He was completely game. (You gotta love Byrd!)

The Big Night Arrives

It was a cold yet surprisingly dry day in Seattle. When Brian arrived at the restaurant, he found James waiting at the bar, dressed to the nines, beside himself with excitement, and well into his second single-malt "nerve tonic." Brian also found Byrd and the film crew in full swing.

We don't know if you have ever seen a film crew in action, but it is pretty amazing. They are kind of like an invading army and the good folks at Dahlia's were completely accommodating. Byrd had brought four cameras and a massive load of sound and lighting gear. When Brian arrived they were hanging a massive rig of lights over a long dinner table.

"You are just going to sit and talk," Byrd said. "Don't worry about us. Just be yourselves. We'll just be in the background filming everything."

"Wow," Brian said. "Okay."

"I know," James smiled, surprisingly relaxed (see aforementioned "nerve tonics"). "It's going to be amazing."

Brian David Johnson gets the Vintage Tomorrows Dinner Party started. Pictured are (Clockwise around table from left) Marshall Hunter, Claire Hummel, Brian David Johnson, Anina Bennett, Paul Guinan, Jordan Bodewell, Thom Becker, Kevin Steil (arriving late), Phil Foglio, Kaja Foglio, James Carrott, Diana Vick, and Martin Armstrong. Notice Donal shooting in the background. (photo courtesy of Byrd McDonald)

James and Brian sat down in a booth away from the action and recorded a little podcast as we prepped for the evening. Then the steampunks arrived and the entire group sat down to dinner.

FIRST COURSE: STEAMPUNK SOUP

What is steampunk? How did you come to it?

The table was packed shoulder to shoulder with steampunks. A grand and glorious sight indeed. "So we have a new version of our joke," Brian said, baiting the entire table.

"A joke?" Diana Vick chimed in quickly.

"The way we have described *Vintage Tomorrows* has always sounded like the beginning of a joke," Brian explained.

"Not a very good joke," James said with a smirk.

"A futurist and a cultural historian walk into a bar..." Brian started then paused. He restarted, "A futurist and a cultural historian walk into a bar and ask: What can steampunk teach us about the future?"

"And what did you learn?" Martin asked.

Paul Guinan, Anina Bennett, and Phil and Kaja Folio arrive to the Vintage Tomorrows Dinner (photo courtesy of Byrd McDonald)

"We'll get to that later." Brian held up his finger stopping Martin and smiled. "But tonight we have a new joke. A cultural historian and a futurist walk into a restaurant and have dinner with a group of steampunks to find out what they can teach us about the future."

"And what can we teach you?" Martin quipped, not easily silenced, even by a master.

Brian quipped back, "That's really what we're trying to get to tonight."

"And to say thank you." James raised his wine glass, defusing with booze.

"Yes! For sure!" Brian added. "We wanted to sit down with you and just talk. We are at a really important moment in time with technology and steampunk and popular culture. But before we begin James and I want to thank you. Without you, this project would have never happened."

"Thank you for all your help and thank you for coming," James added, standing up. "Cheers!"

The table was a flurry of wine glasses clinking and people smiling at each other from under top hats, bowler hats, and perfect, coiffed hair. The appetizers flowed in. It was going to be a good night.

"Thank you for inviting us," Anina said, retaking her seat.

"So what is steampunk?" Brian began. "I know there's a lot of robust conversations around that simple question but let's go around the table and see what steampunk is to you and how did you come to it? Let's start with you. Martin."

"Oh dear, I have to go first," Martin scanned the table with a devilish smile. He's that kind of guy, smart and sly. Martin also has presence. He's a big guy with a booming voice and a thick full beard. There's a mischievous twinkle in his eye that is both wildly intelligent and sometimes worrying. You never know what he is going to say but you know it's going to be sharp and usually pretty darn funny.

Robert Martin Armstrong is the co-founder, along with his wife Diana Vick, of Steamcon. Diana was seated to his right. Steamcon is a steampunk convention that they throw every year in the Seattle area with a small army of people helping out. Steamcon wasn't the first steampunk convention, but it came along early in the new subculture boom and has become a mecca that draws steampunk fans from around the world. We'll spend some notable time there with James later in the book.

"For me steampunk is where all the major creative people have suddenly found themselves," Martin began, making sure to look at everyone all around the table. "We are reimagining the Victorians' original views of the future. Their science fiction has become our science fact. Steampunk is a group of people looking back to look forward again and say: What if?"

Martin looked directly at Brian (ask Martin for his opinion, and you get it—both barrels) and continued. "I always loved that question: What if? What drew me to steampunk was that most science fiction nowadays is all cautionary tales. But the Victorians always thought that science will save us. 'We'll have the cures to all diseases! We'll go to the planets!' It was always positive and upbeat. If I have to pick a place to hitch my truck to, then it's to *upbeatness* and creativity."

"That's one of the things we've seen in our research," Brian replied. "It's lovely that steampunk is about optimism. It looks back to a time where instead of people looking at technology in such a negative way that it was incredibly optimistic."

"I love the storytelling aspect of steampunk," Diana said. Then she added: "Most people at this table will tell you I'm extremely opinionated."

The entire table chuckled, ranging from friendly, knowing chuckle to, well, a little more tense. Diana surveyed the table like she owned it. Her bearing was regal and her fashion pitch-perfect. It was clear just from her poise that she took steampunk culture *really seriously*.

Diana has definite opinions about what *is* and what *is not* steampunk. At the time of the dinner she had just come out with a book called *Steampunk Archetypes*. With such a strong point of view, Diana created friction. But she also gives you a pretty strong impression that at least some of that friction is not entirely without intent. Strong voices and good arguments are essential elements of a healthy culture, and Diana seems to know this intuitively. If you ask Diana about her role in the subculture, she'll tell you she sees herself as an educator for all things steampunk. That's true, but she's also a catalyst, spawning conversation whereever she goes. It's not surprising she's married to Martin.

"I consider myself a bit of a purist," Diana finished.

"Oh! I'm the opposite. I'm not a purist at all," Kaja blurted out then smiled broadly. Everything about Kaja Foglio is direct and wide-open. She broadcasts an intense childlike wonder—but a wonder tempered by the experience of a person who has lived her life in her own way for a long time. Kaja's going to be Kaja, the rest of the world be damned! In 2002, Kaja, along with her husband Phil [seated to her right], created the wildly popular comic book turned web comic series Girl Genius (*http://girlgeniusonline.com*). The comic is, to put it mildly, a hoot. Its tag line is: "Adventure, Romance, MAD SCIENCE!" We consider it mandatory reading; good thing it's been online since 2005. Kaja and Phil are well-known—both in and outside of steampunk—and have fans all over the world.

Kaja continued, "I look around at all this wonderful stuff people are making and say: 'This is all wonderful and I love it and I'm really happy to see it. I don't get too picky.'"

"That's something else we found," James spoke up, pushing his glasses back up on his nose. He's not always aware of what a professorial caricature he can be. The fact that he's not actually a professor makes it even more funny. Must be a historian thing. "Steampunk is a culture, a subculture. From a historian's point of view, it's become a broader movement. And that has been facilitated by the Internet. That's part of the fun of it. You have people coming together through all these really contemporary communication devices to talk about rewriting history and how they prefer to type only on manual typewriters."

"It's also about finding people that are like you," Anina added. "That's how subcultures happen today." Out of everyone at the table (Brian the futurist notwithstanding) Anina Bennett was the least "steampunk." She'll tell you that she's more "punk" than "steam," and while that's true, she's also a sharp-witted and savvy historian. Starting in 2000, Anina and her husband Paul Guinan (seated to her right) created *Boilerplate*—a huge critical and financial success, but more about that later (James considers it better history than most "real" history he's seen—and that's saying something). Anina always had something insightful to add to any topic we came across. When she spoke you could tell she'd really thought through what she was saying and came from a place of strength and expertise.

"We're a bunch of tech-savvy Luddites," Marshall laughed loudly and it quickly spread to the rest of the table.

"Kaja, how do you define steampunk?" Brian asked once the table had settled down.

"I try not to define steampunk." Kaja quipped back, scanning the table with a wry grin and quickly following with, "I don't like it when people shout at me." The table erupted again and the waiters rushed to fill the quickly emptying wine glasses. "But really," Kaja smiled. "I got into steampunk because it has pretty colors."

"Oh Kaja, we're not allowed to say that," her husband Phil gave her a hard time.

"But it's true," Kaja held up her hands, a helpless servant of the truth. "Before we started going to steampunk conventions my friend Jilly said to me, 'Oh you should go to this thing called Etsy and do a search for steampunk.' So I did and there were eight pages of stuff... Now there are several thousand. At that time it was small and people were talking about Steam fashion. I looked at the pictures on the website and I thought, 'Good heavens, these people stole my clothes from high school. How strange.'" The table laughed again and Kaja kept going, "Then I went to a steampunk convention down in the Bay Area and I was in. We discovered the whole subculture part of it there. It was a really cool surprise."

"I think steampunk is catching on and becoming so popular because it's science fiction unshackled from all that tedious reality stuff," Phil added, with a dismissive flair that made his perspective on "tedious reality" quite clear. Phil Foglio is the artistic other half of *Girl Genius*. Everything about Phil is big—his body, his voice, and his ideas. He's a caricaturist who's a character in his own right. "When the Victorians wrote, they said, 'Oh this is great. Let's build a big gun and shoot a ship at the moon!'"

"If we really did that the people inside would be paste when they got to the moon. But we don't have to deal with that in steampunk." Phil was clearly winding himself up for a tale. "Steampunk takes a more optimistic view of technology. It's more concerned with the things that people can imagine as opposed to why we can't do something. So much science and science fiction today says you can't really do this and here's why. That's just depressing! Nobody wants to read that anymore."

As Phil spoke, most of the table nodded along with him, fueling his fire. You got the sense that this was steampunk gospel straight from the apostle's mouth.

Lastwear founder Thom Becker speaks at the Vintage Tomorrows dinner. (photo courtesy of Byrd McDonald)

"Steampunk has a sense of fun. People like that. For the last 50 years all the old famous science fiction authors have been telling us why computers are going to destroy us all. It's gotten to the point that people feel like they can only use an old typewriter to do anything because they're afraid their house will explode."

The entire ta ble laughed as Phil settled back into his chair with a smile and a glance to Kaja. Both Foglios can command a room—and do so with a refreshingly bold and insightful wit.

"When I started off designing clothes, I'd never heard of steampunk." Speaking up from the far end of the table, Thom Becker seemed long and thin and spoke with a British clip to his voice. He was a designer and a tailor and his precise diction seemed to come from the same place as his meticulously designed clothes. His open source design label, Lastwear, is known in the steampunk world for high end, sought-after clothes. It's also quite safe to say that Thom is the most dapper human being you could possibly meet in this day and age (and in many others). "I had an interest in the [Meiji] Restoration Period," he continued. "I designed clothes that crossed traditional period styles with traditional Japanese clothes. When I started showing my designs everyone kept saying, "We love your steampunk clothes."

"I didn't know what they were talking about." Thom motions to his own wardrobe, a crisp blend of Victorian England and Japanese samurai. "It makes sense to me now. So much of steampunk is about telling stories. That's why I got into design in the first place. I like using narrative as a way of understanding human psychology. Steampunk really allowed me to open up and tell stories. The entire scene has been graciously welcoming to me."

"Why do you think steampunk culture was so welcoming to you?" Brian asked.

"I've been a big fan of the open source movement in software," Thom answered, leaning forward on the table in elegant emphasis. "I've always wanted to bring open source software into the real world with physical objects. So, from the very beginning I had the notion that all the fashions we were designing would be licensed under Creative Commons (see Chapter 3). That would allow people to do whatever they want with our designs. Steampunk is DIY (Do-It-Yourself, more on this in Chapter 10 and Chapter 11). It's about creating your own persona, your own character. That's really different and special. If you go to a different gaming or science fiction convention you'll see people dressed up as other people and other characters. But in steampunk there really are no accepted characters. There's no one to dress up like. You are your own character. You get to write your own story."

"That's true." Martin shook his head in faux disgust. "I often get asked who I'm dressed as."

"Right!" Thom replied, holding up his hands. "Happens to me all the time! I'm dressed as *me*!"

"There's a definitive narrative to your clothing," Anina said to Thom. "I hadn't thought about it that way before, but I remember the first time we saw your designs at Steamcon in the fashion show. They really stood out. They are different than everything else. They had their story... and they were beautifully designed," she added with a nod.

"One of the things people don't realize about our fashion designs is that they all have a huge narrative back story." Thom smiled in reply to Anina, warming to the resonance of his ideas. "I'm always imagining ways to mix and remix cultures and encouraging other people to do it with our designs as well."

"That's what I love about it." Anina pointed at Thom's immaculately designed top coat. "There's so much creativity when people create their own characters. It's a form of storytelling. I love that. We've gone to tons of comic book conventions. When people dress up there, they dress as other people's characters... which is creative in its own way," Anina added, not wanting to offend any comic book convention fans who might be at the table. "But in steampunk I love to see that creative energy being put towards creating your own characters."

"Let's keep moving around the table," Brian jumped in. "Jordan?"

SeapiaChord founder Jordan Bodewell speaks at the Vintage Tomorrows dinner. (photo courtesy of Byrd McDonald)

"Steampunk is an aesthetic movement based on Victorian science fiction," Jordan answered frankly. "But really I came to steampunk because of *Chitty Chitty Bang Bang*".

"When I was a kid growing up in the middle of nowhere in the 1970s *Chitty Chitty Bang Bang* made me think that the past was filled with magic," Jordan kept going.

Jordon Block is a promoter and aficionado of steampunk music. Like many in the subculture, he also goes by an assumed steampunk moniker. He turned Jordan Block into optimistic avatar Jordan Bodewell. His music label SepiaChord is known for well-cultivated steampunk compilations and his website is the place to go for reviews and news. You could tell that Jordan had an edge to him, too. He was like a double-sided coin, capable of flipping from super-cynical to wildly optimistic. You might at first think this flip was Block to Bodewell and back again, but there was more to it than that—Jordan's threads run deep. When Jordan cares about something, he clearly cares deeply. Right now, he cares about *Chitty Chitty Bang Bang*.

"It was all technology-based," he continued. "That race car should have died but all you had to do is slap a boat on the back of it and then of course it could fly! 'Cuz that's how things work in the past. It's magic. For me steampunk is about rescuing that past and making things that reimagine that past."

Boilerplate creator Paul Guinan speaks at the Vintage Tomorrows dinner. (photo courtesy of Byrd McDonald)

"That's great!" Brian burst out, clearly a fan of *Chitty Chitty Bang Bang*. "How about you, Paul?"

Paul answered after a thoughtful pause. "My definition of steampunk is fairly straightforward. It's Victorian science fiction." Paul Guinan's voice was easy but metered.

Paul is the visual side of *Boilerplate*, complementing his wife Anina's verbal grace with a meticulously creative aesthetic. Paul chooses his words carefully, giving a sense that he's deeply conscious of his audience and wants to be precisely understood. In this way, he sounded a little like a professor when he first started talking, but the kind whose classes would have long waiting lists.

"As a history buff with a bit of a military bent, I define steampunk as starting with the [American] Civil War and ending with World War I. I came to it at a very early age. My parents were both artists and my father was also a history buff so that helped. I also found science fiction as a kid watching shows like *Star Trek*. The idea of combining science fiction with history was natural for me."

"Oddly enough *Chitty Chitty Bang Bang* was another formative movie experience for me." Paul smiled at Jordan. "Another good movie for me was *Westworld*".

"What I loved about *Westworld* was the mash up of science fiction and robots with Wild West," Paul explained. "That blew my little kid's mind! That's what got

Boilerplate overlooking the White City, 1893 World's Columbian Exposition (image courtesy of Paul Guinan)

me into steampunk and spawned our book *Boilerplate*. I liked telling a historical story, set with all the wonderful historic trappings but including an element of science fiction. I realized that if I included a robot like in *Westworld* and had him play the guide through history then it could be something really interesting."

"The secret of *Boilerplate* is that it's a history book with a robot in it," Anina added.

"You know what?" Jordan interrupted. "*Boilerplate* has been my gateway drug for new steampunk for people.Typically when I tell people about steampunk they think it's only about goggles and gears. But when I pull *Boilerplate* off the shelf and they look through it, they are immediately sucked in."

We hadn't even gotten all the way around the table yet! Brian kept things moving: "Anina, how do you define steampunk?"

"Steampunk is a subculture; a creative community and aesthetic movement that grew out of a science fiction subgenre that was inspired by Victorian science fiction and alternate history," Anina explained in elegantly meticulous detail. She may be a "punk," but don't be misled—this woman has a way with words. "I came to it through working on *Boilerplate*. Like everyone else at this table, I love things that are beautifully made and finely crafted. I love things that are both functional

Illustrator Claire Hummel arrives at the Vintage Tomorrows dinner. (photo courtesy of Byrd McDonald)

and ornamental. I love the creativity and the imagination that go into the things that the people in the steampunk community do. Also I was a punk rocker in the '80s with a bit of a Gothy edge," she added with a smile. "So the whole Victorian punk thing combined really spoke to me."

Brian turned to Claire and asked, "How about you?"

"This might sound like I'm trying to play nicely with others," Claire smiled. "But I agree with everyone else."

Claire Hummel is an artistic prodigy. Her art and designs are deeply researched and incredibly smart. Though you might not guess it at first glance, she's also a serious steampunk veteran with strong opinions on history and aesthetics. She may have been the youngest person at the table, but don't for a second let that fool you. She's earned her stripes.

"I came to steampunk pretty naturally," she continued. "I grew up visiting lots of historical places. I remember I wanted to live in museums. I wanted to absorb all that history and culture. At the same time, my dad would drag my entire family to science fiction conventions. The whole family got obsessed with tracking down and meeting the famous actors and writers. Those two things combined naturally led me to steampunk. I really love the Victorian sense of exploration, never giving up on discovering new things and new worlds. We may have covered most of the planet but we can still discover new deep sea creatures or go into outer space. That's why I like steampunk—at its core it's about discovery and wonder."

Marshall sat up straight and looked around the table. "And last but not least," he smiled. No wallflowers in this crowd—particularly not Marshall. "For me, steampunk is a more general term than most. It doesn't necessarily have to be Victorian, Edwardian or anything else. I'm not interested in historical reenacting... been there, done that. We're not re-enactors or re-creationists... we're re-imaginers. For me steampunk is fun."

You can tell that Marshall Hunter is a troublemaker just by looking at him. He's a maker and live action roleplaying game (LARP) designer who is passionate about creating and collecting the artifacts and stuff from our collective past. In his basement you'll find an arsenal of historically accurate weapons (one of the first conversations he and James ever had was about live-firing 1860s blackpowder rifles) and a treasure trove of bizarre knick-knacks. It's a crazy personal museum that's hard to wrap your head around. But underneath it all, Marshall is a troublemaker plain and simple. He was in good company at this table.

"The steampunk community is this incredibly creative dynamic group of writers and builders and makers," Marshall continued. "I'm a maker. For me, steampunk is all about making." A number of his creations decked out the room we were sitting in, giving our back room at Dahlia Lounge an added kick into the alternative past.

"I agree," Jordan added. "I'll say it's maker-based and I'm not a maker. Sure, I have a costume design degree and I can make you stuff but I am not an imaginative person. People say that making is the religion of steampunk."

"Absolutely!" Paul blurted out, slapping his hand on the table. "If it weren't for my ability to build a little robot figurine, I would not have been able to tell the story of this character in the way that I did. *Boilerplate* would have never happened. I had to physically make the figure first before I could put it into history."

Either the wine was kicking in, or we were seriously on to something. Both, probably.

"A lot of people make things but many times we're also remaking things," Martin added. "When you remake something you put your personal vision into it. That could be a robot," he pointed to Paul, "or fashion," he pointed at Tom. "Or even just reimagining history, but it lets people see it through your eyes."

"I think Marshall nailed it in that we are not reenactors, we are reimaginers." Jordan said to the table.

"Wait a minute," Kaja held up her hands. "You guys talk about the past and how wonderful it was... but it wasn't so wonderful for everybody. One of the things I like about steampunk is you can reimagine it so that it's better. Let's reimagine a past that could have been fun for a lot of different kinds of people, not just the rich upper class white male Victorians."

"That is one of the things we're doing." Diana continued the thought. "We are redressing some of the wounds that were made by Victorian society. People like to say we are glossing over the horrible things that the Victorians did but I don't agree. We're actually making up for what happened, making a better past.

"There's a steampunk I know," Diana continued, echoing, along with Kaja, some of the sentiments Cherie Priest expressed in Chapter 4. "He's an African American gentleman and the reason that he's in steampunk is because his ancestor couldn't have done it. For him, he's making up these stories; making up a better past that really empowers him to do some fantastical things."

"But really there are no universals in steampunk. I don't think steampunk is universally optimistic," Anina said. "I don't think it's universally just fantasy and not fact based. The work that Paul and I have done with *Boilerplate* is an exception to both those points."

The table erupted into argument; some agreeing with Anina, others violently disagreeing. Well, spoons were gesticulated and voices were raised. A "violent argument" among a group of steampunks is a lot more cricket than NFL. "BUT wait!" Anina stops them. "That's what I enjoy about steampunk. There's room for *all* those things."

Another knockout point. We may not all agree on the answers, but we clearly had interesting questions in common around that table.

"I don't think it's utopian or dystopian." Marshall sat up. "It's about making. For too long we've lived in a black box society. We grew up not knowing how our VCRs worked. Now we have a group of people who when they see a black box or a VCR or any kind of closed technology, the first thing they do is grab a hammer and a screw driver and say, I'm gonna know how this works."

"One of the key things about steampunk and the appeal of the Victorian era is that it was the last point in history when a high school education could give you enough knowledge to do amazing things," Thom stated.

"Yes!" Phil yelled, nearly standing up. "It's the idea of the mad scientist inventor, the lone guy in the castle making something that will change the world!" Phil looked more than a little like a mad scientist himself as he delivered this pronouncement. "It was an individualistic thing and steampunk allows people to play out that story in their head. They could make something amazing in their basement. It doesn't have to come from some corporate lab."

"I think when we say optimism we don't necessarily mean that we're all happy cheerful all the time." Diana spoke slowly and wanted to make sure that she was understood. "We're saying that today's science fiction is so cautionary. There's so little science fiction that talks about exploration and invention. Modern science fiction tells us: 'Oh God, don't go build giant robots. They'll kill us all!' But Victorian science fiction says: 'Yay! Let's go build giant robots! Oh shoot, they killed us.'"

The entire table roared with laughter. Brilliance. We had no doubt that line would end up in the film.

"It's a spirit of adventure and curiosity," Kaja added when the room quieted. "If we ask ourselves: What does this button do? The answer is: Let's push it and find out!"

Even more laughter than before. Yes, the wine was taking its glorious toll, but we'd really come together as a group around some really key insights. And heck yea, we'd all press that button. The red one first.

"Steampunk is a rebellion against the entire industrial design aesthetic of the last 50 years," Phil continued. "High-tech designers today are basically trying to boil everything down into a magic box, like the iPhone or the iPad."

"It's beige, you'll love it!" Kaja added. The husband and wife team were on a roll.

"It has no moving parts!" Phil did his best carnival barker impersonation. "How does it work? You have no way of finding out!"

"Get me my hammer and screw driver!" Marshall yelled out. "I'll find out how this baby works!"

The room broke out into laughter as the first course ended. We'd come a long way toward understanding steampunk in a pretty short period of time. We may also have found a darn good stage comedy act or two in the bargain.

We stood and stretched, took the necessary human breaks, and gave each other grief about what had been said and gone unsaid. Brian, Byrd, and James conferred briefly, exchanging knowing grins. We really had something here. And what's more, it was only the beginning.

MAIN COURSE: THE MEAT OF THE MATTER

Why now?

"Welcome back everybody to our second course," Brian began. "This is where it gets good. This is where it gets dicey and we start to argue. I encourage you all to wildly interrupt each other."

Quizzical glances were exchanged across the table. Er... Brian had been with us during the last course, right? Of course he had. He was just upping the ante. What good's a fire if you can't splash on a bit of gasoline?

"Actually Brian," Martin interrupted with his devilish smile. "You do realize that steampunk is much more civilized than that. We have our Victorian manners so we probably won't be interrupting each other."

"Somehow I doubt that," Brian replied while the table chuckled. "But, before we begin, we have a new person at the table."

The table applauded as Kevin took his seat.

"Hello everyone, I'm Kevin Steil," he waved modestly. "I'm the Airship Ambassador. I write the Airship Ambassador website and blog."

Kevin is soft spoken, gentle, polite, and kind to everyone. That's what makes him the perfect ambassador for steampunk. He is one of the best-read, best-connected people in the subculture. He's a great example of a key difference between steampunk and its bohemian antecedents—Kevin is all about sharing knowledge and welcoming people in. You just can't find anyone more approachable and easy to talk to. His huge expertise is dedicated not to policing the boundaries of who gets to be a steampunk, nor is it about self-aggrandizement—it's about sharing.

Plus, and this can't be overstated—the man always has the best damn coat in the room. James and Kevin hadn't actually met in person before this dinner, but James later confessed that in all of his research, he'd never seen anyone who could out-class the Airship Ambassador's coats. No small feat in a flamboyant world of costume *par excellence.*

"So we have an ambassador at the table," Brian nods to Kevin. "Okay, so the next part of the dinner is a bit different, but please eat and drink." He motioned to the wine. "Don't forget the wine."

The wine had been serving us well, after all, and needed to be given its due.

"Let's look at where we are today. One of the things James saw in our research was that if you map the cultural activity around steampunk over the past few years you see that there's this huge spike in activity," Brian mimicked the sharp

Airship Ambassador Kevin Steil arrives at the Vintage Tomorrows dinner. (photo courtesy of Byrd McDonald)

curve of the graph with this hand. "If you collect all the steampunk *stuff* that's being produced, that's TV and games and comics and books and music and all of this stuff, there has been an incredible spike in cultural activity around steampunk." The whole table nods while he speaks. No surprises there—they'd all seen the explosion up close and personal.

"Now my question is: *Why?*" Brian said.

"It's how cultural change happens," James added, unable to contain his excitement on the topic. "There is something particular to steampunk that's aware of the historical patterns that's unique. It's very specific to technology." James pushed his glasses back up on his nose again (apparently not to be out-professed by the esteemed Mr. Guinan) and continued. "Today people want to know how their technology works. Hacking and making is becoming more mainstream. People want to break apart a physical object and look inside and make it into something new.

"I see writers and creators doing that with history as well," James continued. He was on a roll. A tipsy historian can be a blessing or a curse, depending on how much you enjoy hearing lots of words. "They are making history into something new. What's exciting to me is that we're not only doing it but we are conscious of doing it. That's very different from other subculture movements I've studied like the Beats and the hippies. Steampunk is consciously remaking the past, remaking the present and remaking culture."

"But *why now?*" Brian added finally, politely shutting down his insightful, but perhaps overly-enthusiastic colleague. James gets inspired, wine or no, and he'll talk your ear off unless you redirect.

"People think that technology is getting good enough that we can now rediscover our ability to make things," Kaja started things off. "We can program our own things and we have the ability to share that knowledge."

"That's the exact point right there!" Martin pointed passionately at Kaja. "It's not only about doing it for ourselves but it's about sharing the knowledge."

"James, you mentioned the beats and the hippies; I hadn't thought about the connection to the '50s and '60s," Anina replied. "I'd never thought about the social change that was happening then in relation to what's happening now. The design sense of the '50s was very clean, very buttoned down, and the '60s really busted that way open. People took it apart and tried to put it back together again. That makes me hopeful that steampunk might be able to do something like that."

Across the table James quietly smiled at this vindication from one of his favorite writers.

"In steampunk do you see the indication that we are in a moment of radical change?" Brian asked, tagging Anina's insight for the affirmation and expansion that it was. Our hypotheses were resonating and growing before our eyes.

"I think there's massive dissatisfaction with the daily grind," Anina answered. "Sure we love technology and it's magical but we don't know how it works and that makes us feel like it's controlling our lives. People think that their lives and our society is out of control. Steampunk is a radical reaction to that."

"Is our culture broken?" James asked across the table. Profound is not actually his middle name, it's Hartley.

"Yes!" Kaja blurted out. "There's a perception that it's broken."

"Absolutely," James smiled at Kaja, obvious that he was a Foglio fan through and through. "When it comes to the latest technology people are asking themselves if they really need it. There are a growing number of people who want more control of their technology. They want to open it. They want to void the warranty. That phrase is even on t-shirts now." James pointed at his chest.

"We don't even *own* it." Anina shrugged her shoulders. "We don't have control of our books or our music."

"But it's not just our devices... it's our own lives." Kaja seemed a little overwhelmed and yet passionate at the same time. "Do you have control over your own life? People are starting to look for alternative ways of living. Maybe I don't want to work in a cubicle. Maybe I want to make a strange little corner for myself in the world."

"We're getting too big for our britches." Jordan slowed down the table with a critical curve ball. "Steampunk isn't even a subculture yet. We are a micro-culture." He paused as if struggling to make sure he was understood. "I do want to put weight on what we're doing but in my heart I'm not sure. When I go to my son's school, no one knows what steampunk is. They don't care. American culture is driven on variety and novelty and steampunk hasn't reached a critical mass."

"I don't think we're putting too much weight on it," Anina replied. "Whether or not steampunk becomes a broad cultural movement is a different question. There's a deeper reason more and more people are being drawn to the aesthetic and the narratives. They are drawn to it for cultural, psychological, and historical reasons."

"Punk rock was a micro-culture," Martin agreed and added. "It exploded all over the mainstream for a while. Then it went to be a small culture again. It's the same kind of thing.The reason steampunk is so much bigger is the Internet. The Internet is a giant magnifying glass."

"But why have so many people picked steampunk?" Brian pushed the table for more detail. He was used to talking about technology. He leaned forward as he spoke, clearly wanting something more—something new.

Anina stepped up, "The aesthetic pendulum swings back and forth all the time. Fashions come and go, from one extreme to another. On the surface that's part of it. But there is something else. There are a lot of people who have this desire to somehow take control of their own lives and their own identity. They want to remake themselves and maybe even society, culture, and politics..."

"Steampunk epitomizes the use of knowledge for exploration," Kevin, the Airship Ambassador broke in hesitantly. "Steampunk gives you freedom to be an individual. Technology today all looks the same, an iPad looks like every other iPad. A Droid phone looks like every other Droid phone. They all look the same. People are reacting against that. They want some individuality. They don't want to follow the crowd."

"Are people rethinking their relationship with technology?" Brian continued to push the table (literally as well as figuratively—it's not just historians who get excited around tables full of good folks, good food, and good wine).

"Yes, I think we are..." Martin started.

But Phil broke in. "It's about understanding. People love steampunk because they understand how the technology works. For the last 50 years we've been designing the workings of technology to be hidden. Everything is smaller and sleeker. Steampunk celebrates being able to look inside; you can see the cogs and the gears. People like to know how things work. They like..."

"...to see moving parts." Paul finished the sentence. Some threads were coming together here—parts of a conversation we'll continue throughout this book.

"You two are talking about two different things," Jordan pointed out. "Understanding and seeing are not the same thing."

"But people *think* they understand it," Phil added. No insignificant nuance in a world where truth is often something we're left to create for ourselves.

"I can't understand a steam engine." Jordan wouldn't let Phil (or anyone else for that matter—god bless him) off the hook. "I just enjoy watching it happen. When I go see a steam train, I don't understand how any of that works. I enjoy watching it happen."

"It promises the idea that you can understand it if you just study it a little," Paul clarified.

"That's interesting," Brian pointed out. "It's not the understanding of it, but it's the promise that you *might* be able to understand it."

"There's a word that hasn't been tossed out yet." Phil smiled and looked up and down the table. "The word is 'empowered'. People want the ability to make what we want for ourselves."

"It goes beyond customization," James agreed. "People want the ability to take their technology apart and put it back together again."

"We need to change our culture to be about selling products because they are unique and special and distinct for you," Jordan replied directly to James. "But that's not how things are valued today."

"This is how culture changes." James smiled like only a cultural historian can smile.

Jordan slapped the table. "That's what I'm saying!"

"This is how culture changes," James continued. "Slowly and not all the way through. The '60s had their impact, not the entire impact they wanted, but it had impact. Then some of the very same people from the '60s went on to have impacts in other areas, not the entire impact they wanted, but it did make change. But that's how culture changes. We are changing culture." Then that smile again.

Director Byrd McDonald captures the Vintage Tomorrows dinner. (photo courtesy of Donal Mosher)

DESSERT: THE FUTURE LOOKS SWEET

What can steampunk teach us about the future?

"It's time for the third course, it's dessert and it's about the future." Brian brought everyone back to the table—an effort not dissimilar from herding a dozen sharply dressed and highly opinionated ferrets. "What can steampunk teach us about the future? If we are experiencing cultural change then what can this subculture teach us about our broader culture?"

"We are talking about making technology more human, making technology itself more human," James added. "In our research we've seen that people want their technology to have a sense of humor, a sense of history, and a sense of humanity. Does that resonate with you folks?"

"I want machines that will keep us from killing each other!" Phil declared.

"That's scary." Kaja cocked an eyebrow at Phil. "That could be a scary scenario... why would you want any technology to have that much control..." She paused and contemplated for a moment. "All I can think of for what I want for the future is a Victorian house with little robots that clean it for me. Now that's exciting!"

The table laughed, a deep laugh that was full of wine and great food.

"A very honest answer," Brian remarked.

"I use my steampunk values at work pretty constantly," Martin said honestly. "Victorians were always looking for the new and the great I want for the future. I want to go to the stars. Steampunk lets me go back to the time when people actually believed that not only was that gonna happen but we were going to have roundtrip tickets to Mars leaving daily from Amsterdam and Copenhagen. That motivated vision of a bigger world, a world that was more united, more cosmopolitan."

"You said humor, history, and humanity," Jordan spoke up right away. He was lost in his own thoughts. "Humor's not the right word. Steampunk has more of a sense of play."

"...laughing with whimsy," James added.

"Let's be clear," Diana broke in, deadly serious. Then she pointed directly at her decolletage. "I'm a woman with an octopus coming off my breast. I have tentacles coming off my bosom!" The table exploded with laughter. "I mean really... do we want people to take us that seriously?" She patted the little octopus on the head. The table continued to laugh. "It's not about being that serious. It is about humor There's nothing wrong with being funny."

"It's whimsy." Marshall smiled.

Paul spoke up, "First and foremost when people look at *Boilerplate*, they smile, they laugh, and that's exactly what I was looking for. That was my intent from the get-go."

"At a certain point," Marshall continued, "if it is so ridiculous then it's so funny that it's engaging."

"There's nothing wrong with funny," Anina said seriously. "People absorb information much better when they're being engaged and entertained."

Paul stopped the entire table with a flourish of his hand—an actual flourish—and began, "George Bernard Shaw said, 'If you're going to make people think, you also better make them laugh, otherwise they'll kill you.'"

That killed everyone. We collectively lost it, then took the opportunity to recover with a crack of Crème brûlée or bite of cake. The professor game was tilting. What Paul lacked in glasses he made up for with erudite wit.

"This is why Jon Stewart and Stephen Colbert are so popular," Jordan commented. "Only the court jester can make fun of the king."

This get everyone at the table talking. (We'll hear more about John Stewart and Stephen Colbert later in the book—those guys have cultural clout, and for good reason.)

"Steampunk is smart humor," Claire said. "It's witticism."

Martin added, "...it's *dry* and witty!"

"Steampunk gives people a venue to be smart about humor," Claire continued.

That killed everyone at the table. (photo courtesy of Byrd McDonald)

"It goes back to James's point about Steampunk and why it really started to become popular in 2005." Martin pointed at James and paused. "Look back at the news stories in 2000, they were bleak and boring. People were looking for more; they wanted something fun? Where can I find whimsy? People just wanted a little escapism. They were still grounded. They didn't want to fly away entirely but at least they wanted a little escapist fun."

"You know there was something interesting I heard on panel at a convention once." Diana stopped everyone. It was obvious she had a story and no one was about to interrupt her. Diana can have that effect. "The people on the panel said that science fiction was a genre that was dying but that steampunk might be the only possible way for science fiction to regain any power. These days science outstrips science fiction so quickly. Most authors can't write about the future because it's happening too quickly. It's really difficult to write hard science fiction right now without being a futurist." Diana smiled at Brian. "You can't envision the future because the future is two minutes away. But with steampunk we can go back and talk about things that didn't happen, couldn't happen, and there's so much more potential in that."

We'll come back to this point later in the book. It's not accidental that William Gibson's most recent novels are set essentially in the present day.

"So the future is no longer the realm of the imagination." James smiled. "It's the past. It's steampunk?"

"I think so." Diana nodded.

The end of an amazing night! Cheers! (photo courtesy of Byrd McDonald)

"I don't think we'll find any better thought to end with," Brian said, standing and raising his glass. "And with that James and I thank you all for coming. Thank you very much. Cheers!"

A Film is Born

What an amazing night! It was late in Seattle. The documentary crew, Byrd, and Brian were staying at the same hotel across the street from Dahlia's. It was 2AM. The shoot was over and all the gear had been packed up. All the steampunks and James, who had finally been permitted to don his top hat (it blocked out whole camera angles) had gone home. The crew, Byrd, and Brian kept the bar open simply by ordering drink after drink.

Everyone was pumped up on adrenaline. It had been an amazing night. The conversation had gone places that were funny and unexpected. Byrd's crew had captured everything. The entire experience had felt comfortable and real—just a bunch of Steampunks, a cultural historian, and a futurist sitting around a table having dinner and talking about the future. It couldn't have been better.

As the bar started to wind down, Byrd came over to Brian, sipping on his second or third beer (who's counting at a time like this!). He leans against the bar and says, "BDJ, that was amazing."

"Yeah, I know," Brian replied. "They are a really good group of people. I thought it was..."

"No," Byrd stopped him. "That was an amazing conversation. I think I'm going to make a documentary about it. I want to make a documentary about *Vintage Tomorrows.*"

"That's great!" Brian held up his glass to an empty bar.

"No," Byrd stopped him again. "I want to make a documentary based on your research and your book before you and James are done with the book. I want to capture the process... I think it will be fascinating."

"That's a great idea," Brian replied, looking around for someone to tell. The bar was empty. The city was asleep.

"What do you think?" Byrd asked.

"I think that's awesome!" Brian replied to the empty bar.

That's how *Vintage Tomorrows* the documentary started in a bar at 2am in Seattle; pretty much the same way the entire project began, just a few blocks north and a couple more east. It sure seems like good things happen over beer on a Seattle night.

A Note from the Historian

"Chronological living is a kind of lie. That's why I don't do it anymore. Existence doesn't have more meaning in one direction than it does in any other."

— CHARLES YU

How to Live Safely in a Science Fictional Universe (2010)

BRIAN AND I *set out from that dinner stuffed full of questions, with just the slightest lingering aftertaste of a few hypotheses. It was time to get down to business—this was a project in earnest. So earnest, in fact, that I left my job at Microsoft and took to the road as an itinerant historian. Talk about a dream job. I immersed myself even deeper in steampunk—talking to everyone I could, reading piles of fiction and watching hours of video, spelunking the web and attending every convention I could reach.*

Let it never be said that a historian's life is devoid of adventure. (photo courtesy of Andy Pischalnikoff)

The following chapters tell the story of this adventure—a voyage that ultimately took me to the other side of the world and then right back to where I started. It also took me beyond (or perhaps beneath) steampunk, to some pretty core questions about our relationships with technology, history, and each other in an increasingly digital world. It's been one hell of a ride. That ride, however, has produced a story that can't really be told in a traditional historical fashion.

So, in keeping with its subject, the tale you are about to hear has been punked. *We're going to jump around in time a bit, and present the truths I uncovered in a manner that highlights the insights themselves rather than the minutia of, for example, what part of what conversation happened when. Where this sort of thing is important, I'll call it out —I am a historian after all, and while I'm not entirely certain that linear time is a defensible way of looking at the past, I have a duty (there's no "historian's oath," but there might as well be) to preserve place and time where they are significant.*

That being said, the following story takes place in numerous cities throughout Australia, Canada, New Zealand, and the United States (California, Florida, Nevada, New York, Oregon, and Washington). The conventions and events referenced are: 2010—Steamcon 2; 2011—Emerald City Comic Con, Norwescon 34, Gearcon, Steamcon 3, and New York Comic Con; 2012—The Key West Literary Seminar, Norwescon 35, the Victoria Steam Expo, Maker Faire San Mateo, Burning Man, and Steamcon 4.

Also, because any book or film is as much or more a product of its own time as that of its subject, I'll state clearly that all of the conversations and experiences chronicled in the following chapters occurred between November 2010 and December 2012. Most interviews were conducted in person (it'll be quite clear when this is or is not the case).

The reason I'm saying all this is that rather than placing myself at the center and telling my story from start to finish, I'm going to follow a path that is true for you, *taking a few liberties with my own personal timeline in order to give you a more meaningful view of steampunk and what it has to say about (and to) our culture. This is a work of nonfiction, but Brian and I decided pretty early on that this tale would be told not to ourselves, or for posterity, but for you, our readers.*

It's About Chickens and Teapots

Our historian does what he does best! James journeys into the heart of steampunk and beyond, following and even tugging at the threads of cultural change. He starts with the steampunk of the matter: just what *is* this steampunk thing, really?

History is a family affair with the Carrott family. James brings his daughters Mimi and Beatrix along as he investigates the details and corners of stempunk culture. Seven year-old Mimi has an insight, and from there James agrees he must press on.

The first order of business is to define: "What is steampunk?" Not an easy task. James tracks down Mike Perschon, the "Steampunk Scholar," to get a firm grounding in what he needs to know, then dives into the world of steampunk conventions. After making his way through some poetry we make our way to Canada for the Victoria Steam Expo where we get a different perspective from the convention's ringmaster, Jordan Stratford.

But it all starts with a cup of chicken tea.

Chicken Tea Is Clucking Good

James H. Carrott (Seattle, WA)

> *I absolutely refuse to believe your perfectly logical explanation. Mine is far more poetical and therefore speaks of a greater truth.*
>
> **— ALGERNON SWINBURNE FROM MARK HODDER**
>
> *Expedition to the Mountains of the Moon* (2012)

STEAMPUNK IS a highly visual creature. Plug the term into a search engine set for images (give audience-appropriate consideration to "safe" settings—where there's pop there's porn) and you'll come up with more costumery, illustration, gadget photos, and, well, stuff than you can take in at one (or many more) sitting(s). I started collecting good images right away and tucking them into a folder for reference. I was still working for Xbox when I began this research, and the release of the game Fable 3, with its hysterically campy steampunk-inspired imagery proved a goldmine.

When I sat down at the end of 2010 to write up my first research report to Brian, I put together a screen-saver for my Apple TV that collected a bunch of my favorite steampunk illustrations: a few clips from Phil and Kaja Foglio's infamous web comic *Girl Genius*, some of Claire Hummel's wizardry, and of course, a bunch of the Fable 3 posters, kindly supplied to me by their art team through my friend Patrick, the game's product manager.

Fable 3

Fable 3 is a video game, developed by Lionhead Studios and published by Microsoft Game Studios. It was released on the Xbox 360 within a week of the fateful beer that started Brian and me on our adventure, and sold millions of copies since (over 10 million for the franchise as a whole—video games are a big business). I love roleplaying games, and this one in particular played a big chord on the gamer piano in my heart. The third in a series (hence "3"), it advanced Fable's epic story chronologically—into an industrial age. The game just drips with brilliant steampunk pastiche and is a hella fun play to boot.

One of those images from Fable 3 is of a chicken in a teapot. "Start Your Day with a Cup of CHICKEN TEA" it proclaims, "It's CLUCKING Tasty!" My kids don't watch much TV at my house, but I left the screensaver on a lot, as it kept me thinking (and laughing) about my research.

"Chicken Tea" from Fable 3 (2010) (image courtesy of Lionhead Studios)

Mimi and Beatrix (then 7 and 3, respectively) were captivated by the pictures. Claire's pug dogs in 19th century military regalia, and of course the ad for Reaver

Industries' AUTOPICK: "Tired of Bogies on your Gloves? Get AUTOPICK; With Custom Tools to Pick: NOSES – EARS – TEETH – SPOTS; Ties Cravats, Plucks Eyebrows, & Waxes Your Moustache! Plus belt attachment to deal with Rectal Blockages!"

The kids made me read it over and again, and it was funnier to them each time. Especially once I'd defined "rectal"—7 and 3 are the right audience for butt jokes. (Butt jokes aside, much of this was over Beatrix's head. She was only 3 after all, and "butts are funny" was enough for her at the time.) Mimi, however, was absolutely entranced by the gadgetry, costumes, and fantastical settings. Her own drawings started to feature funny hats and goggles, big pouffy dresses, and teapots. I was bowled over by how infectious the steampunk aesthetic really was. We'd talked about my research project a little, but at 7, the past itself takes some time to sink in, let alone the idea of playing with it. But the funny, insanely imaginative pictures? They rocked her world.

A few months later, when *Vintage Tomorrows* was ramping up into a bigger part of my life, Mimi sat next to me on the couch, scribbling intensely in one of her many sketchbooks. I looked up from my laptop.

"Steamplant" costume/model/photoshop by D. Vick (2011) (photo courtesy of R. "Martin" Armstrong) (left); "Steampunk Queen" by Mimi Carrott (2011) (right)

"Hey Meems," I said (Mimi is a young lady of many nicknames, though she is formally "Amelia" after the renowned aviatrix), "you're still drawing a lot of steampunk-looking stuff. That's cool."

"Yeah." She said, barely looking up. "It's the awesomest."

I'd just shared Diana Vick's Steampunk Archetypes booklet with her, and she was madly costuming heroines accessorized with brilliant old goggles, ray guns, and wrenches.

"Well," I said, ever the over-educating dad, "do you get it? Steampunk, I mean."

"Yep."

Wow, she had one on me there.

"Okay." I subtly slipped into interview mode. "So when I say 'steampunk', what is it you think about?"

Always intense and attentive, Mimi thinks hard about her answer when asked a question like this. Completely straight-faced, she responded as if in class:

"Daddy, it's about chickens and teapots."

I tried hard to hold it in, but just couldn't, and burst out in a fit of laughter. Mimi laughed a little too, kinda nervously until I said, "You know what, Meema. You're right. You're really, really right."

Though Diana Vick still rolls her eyes every time I say this, Mimi really was dead-on. Steampunk *is* about chickens and teapots. She'd articulated something my overly-analytical adult brain couldn't. A child's association of steam with teapots makes literal sense, but the core of the insight is about play—something Cherie Priest talked about back in Chapter 4: "*if you're not having fun, you're doing it wrong.*" Steampunk is about messing with the world around you, putting unlikely things together for the heck of it, and laughing about it all. There are serious sides, to be sure, and we'll get into those later in the book, but it's the sense of inclusive play that "chickens and teapots" implies that's really at the heart of steampunk's challenge to our modern world. It also may well be the key to a better future for us all.

Steamcon and the Steampunk Sandbox

> *I shall not today attempt further to define the kinds of material I understand to be embraced within that shorthand description ("hard-core pornography"); and perhaps I could never succeed in intelligibly doing so. But I know it when I see it, and the motion picture involved in this case is not that.*
>
> **— SUPREME COURT JUSTICE POTTER STEWART (1964)**

While Mimi was dead right, I still needed to be able to draw some lines, no matter how rough—the boundaries of the sandbox, as it were. In order to study something, you first have to be able to articulate what it is. In order to put together the "cultural thermometer" we saw in Chapter 1, I needed a functional definition. The more formal results of this endeavor are laid out with more detail in the *Historian's Notebook* (see Prologue), for those of you with an academic bent. This chapter (well, this book, really) is more about exploration and discovery than social science and analysis. We're chronicling the flight, not the airline's scheduling infrastructure.

I've learned to accept serendipity as a part my cultural research. (Irony too—I am, after all, a historian studying how playing with the past in the present might impact the future.) Some things just drop into your lap and you go with them. As soon as I set down my phone after signing on to undertake this project with Brian, I hopped on the internet and started searching. In order to figure out just what the heck this thing was, I had to get to know steampunk better *in person*. I knew that Seattle had an active steampunk scene—the Seattle Steamrats had a weekly "Steam Vent" that I'd attended once or twice with an artist I used to date—and there was bound to be something going on soon. It turned out that "something going on soon" was one of America's headline steampunk conventions: Steamcon.

The timing couldn't have been better. In just a couple of weeks, the convention would be happening in Sea-Tac, the airport non-suburb south of the city and north of Tacoma (which as a displaced Chicagoan, I tend to think of as Seattle's Gary, Indiana, just without the haunting landscape of desolate mills or the snappy show tune).

I wasn't entirely sure what to expect walking into Steamcon in November 2010. My research had begun in earnest and I clearly had to be there, prepared to take in anything and everything I could. I hadn't planned ahead as well as I might, and had to wait in line to purchase tickets on the opening day of the con. Not a bad thing, as it turned out, because it gave me the opportunity to strike up a couple conversations and afforded the time to just take in the spectacle of it all. I struck up conversations in line and was surprised to meet a woman from Germany who was in LA visiting a friend and came up to check out the con, there being nothing of the sort in her neck of the (black) woods. I met parents who'd brought young kids (complete with toddler goggles), teens, folks of my (distinctly-middle) age, and an older couple sporting an iPad.

It was a long time in line, as I'd missed the pre-registration window, but once I'd obtained my badge and convention guidebook, I was ready to go. But go where?

I'd come to the convention to answer what I thought was a simple question: What is steampunk? I certainly wasn't blind to it, but it was hard to pin down. I've heard more than one person fall back on the old Supreme Court obscenity line:

"I know it when I see it." But I needed something more than that. I'm a scholar, after all—I wanted a definition: something I could use to define the boundaries of what I was studying. Turns out, pretty much everybody else there was asking the same question. Folks shopped around their pet theories, debated in the halls over coffee and the high-end-ish hotel conference water cooler substitutes (you know, those tables with a big shiny dispenser and rows of actual glass glasses). Steam gun-toting cowboys argued Verne and Wells with zombie mechano-cyborgs in bustle dresses.

I met David Maliki ! (yes, he spells his name with an exclamation point; it's not uncalled for) in the vendor area, hocking some of the coolest steampunk stuff I'd ever seen. His webcomic, *Wondermark*, and related merchandise (stickers, posters, t-shirts, and more) struck me as the epitome of steampunk—clever contemporary play with the stuff of the 19th century past, commenting on technology, change, and culture. When we spoke briefly at Emerald City Comic Con the following year, I was surprised to hear that he didn't really think of his work as steampunk. My encounter with *Wondermark* served to further underline my growing suspicion that this thing I'm studying is considerably more than first meets the eye.

Wondermark: Tinkering with Wonder

David Maliki ! launched Wondermark (*http://wondermark.com/*) onto the web in 2003, four years before he'd even come across this phenomenon called "steampunk" on Boing-Boing (as if we needed further evidence that Cory Doctorow is one of steampunk's evangelist "Ginsbergs"). His reluctance to assume the label is as much a factor of his respect for steampunk culture—he declines communion like a conscientious non-believer—as it is an artistic desire to avoid labels as an artist. David's play with the Victorian past takes many directions, tacking in clever zigs and zags all around the cultural landscape.

Wondermark applies a maker mentality to comics, tinkering found images into new works of art. David sells his wares online, as well as at a variety of conventions. For obvious reasons, his work appeals strongly to a steampunk audience, but he finds that his strongest sales come out of places like Maker Faire. In some ways, this ought to come as no surprise. David is the son of a mechanic, raised to tinker with everything from engine parts to clip art. His success in the maker community, however, speaks to a connection we will explore throughout this book—playing with history and culture shares a deep kinship with playing with physical objects. Both are hands-on ways of making sense of a changing world. "As technology grows more opaque," he told me, "people who want to understand something have to do it themselves."

Steam Powered, (image courtesy of David Maliki !) Since Steamcon 2, this clever graphic has adorned my laptop, iPad, and favorite t-shirt.

I continued strolling and discovered that walking the halls at Steamcon, you're as apt to overhear "holy shit, how did you make that thing?" as "dude, that's dieselpunk, not steampunk—totally different" as "yes, these are my real boobs." Conversation wanders from the deeply literary to quirkily flirty in the blink of an eye and it's hard to make sense of it all until you realize what it really is that all these people have in common. *It's not answers that they share, it's questions.*

Wondermark's "Tinker Rules" (image courtesy of David Maliki !) Yes, this is foreshadowing.

My friend Mark Cohan is a sociologist who began his own research into steampunk at this very same con. We've easily dropped gallons of coffee since chatting about these questions. He's come to think of this—questions rather than answers—as part of the glue that holds the subculture together. Remember our quick definitions in Chapter 1? Subcultures define themselves by drawing lines between themselves and mainstream culture. Otherwise, why call it a subculture? But Steamcon's many panels and conversations exposed a tricky bit. Steampunk really seems to be founded on questions rather than answers—big fuzzy grey areas rather than the stark lines more common to subcultures. One of the things Mark muses on is steampunk's "rhetoric of radical-inclusivity," which is, in academic speak, essentially an echo of Cherie's invitation to anyone and everyone to come and play in her steampunk world. In steampunk culture, exceptions to the rule are kinda the rule in themselves. I can't speak for Mark, but from where I sit, it strikes me that a good deal of what he's uncovering is that steampunk is hacking subculture itself. What a fantastic twist!

The overall effect of Steamcon felt something along the lines of twenty-first century software running on nineteenth century hardware. It was all clodgy, historically-inspired, and, to me, eerily comfortable. At the same time, it was cutting edge contemporary. And I mean the bleeding cutting edge—these folks were hacking *everything*. They hacked their clothes, notebooks, display tables, haircuts... even the couple with the iPad had it gorgeously encased in wood, with intricate brass filigree. Granted, this is hardly hacking the iPad itself, but the intricacy and craftsmanship of that case graduated it far beyond "glue some gears on it" in my book.

Just Glue Some Gears on It

"Just glue some gears on it" is a reference to steampunk authenticity—or rather the lack thereof. Sir Reginald Pikedevant Esquire's "Just Glue Some Gears on It and Call It Steampunk" first appeared on the internet in late 2011.

It's uproariously funny and surprisingly insightful. Here's a little excerpt:

Just glue some gears on it, and call it steampunk;

That's the trendy fashion nowadays!

A copper-painted chunk of some nineteen-eighties junk

Will fetch a pretty penny on eBay!

But it got even worse than that, I'm afraid:

And as I went on, I became more dismayed;

For they often didn't modify the things they wanted me to buy

Heavens to Besty!

They'll end up on Regretsy!

Steampunk is more than a mere mode of dress

or decoration, so it brings me some distress

To see a metal cicada described with that name

— SIR REGINALD PIKEDEVANT ESQUIRE

"Just Glue Some Gears on It and Call It Steampunk" (2011)

Regretsy (a website self-described as "Where DIY Meets WTF") by the way, has a whole category titled "Not Remotely Steampunk (*http://www.regretsy.com/category/not-remotely-steampunk/*)."

What's particularly brilliant about the lyrics above is that they make a distinction between stuff that displays effort (even if it *is* "just glued on") and stuff that's just been given a trendy label. It's not black and white—good steampunk or not steampunk at all—but shades of grey. Sir Reginald grants a bit of respect to those that at least try. Inclusivity in action.

"They just glued some gears on it" has become steampunk colloquialism—a way of calling out stuff that lacks heart, effort, or imagination. It's also a great example of a counterculture talking back to its appropriators. This kind of two-way dialogue may well keep steampunk's edge sharp much longer than its predecessors. Counter-appropriation via internet meme, as it were. Ain't no "beatniks" here.

THE STEAMPUNK SCHOLAR

> *Anachronistic electricity, "KEEP OUT" signs, aggressive stares... has someone been peeking at my Christmas list?*
>
> **— THE DOCTOR**
>
> "A Town Called Mercy" from *Doctor Who* (2012)

You can't really *study* steampunk without crossing paths with Mike Perschon, The Steampunk Scholar—a literature professor who's pretty much the definitive guy when it comes to steampunk fiction. I saw his talk, "Steampunk: Technofantasy in a Neo-Victorian Retrofuture", on my event schedule and wondered if Christmas had come early this year. Boy was I not sorry I prioritized that one. If you haven't already, this is the point at which I expect the rest of you to wince and wonder at my sanity. "Technofantasy?" "Retrofuture?" "Okay, 'Neo-Victorian' kinda makes sense, but seriously, why would you do that to yourself?"

Well, a couple reasons. Good ones. First, neither I nor steampunk are afraid of big words, or weird words for that matter. I'm not implying that *you* are scared of any word, big or small, just acknowledging the fact that academic-y sounding stuff can be a big turnoff to a lot of folks. And for good reason—there's a LOT of deeply boring academic work out there, even on the most interesting topics (browse a "cultural studies" section somewhere if you doubt me... that stuff ought to be *fascinating* by its very nature but even I can hardly get through half of it). This all said, in a steampunk context, the fact that something doesn't initially make sense is usually a darn *good* sign that it'll be hella interesting. Good academics (i.e., the appropriately weird ones) know how to make this kinda stuff into one incredible show. You speak to your crowd, and boy does Mike ever do that. Seriously entertaining and informative. I was hooked from his first sentence.

When they've been in the scene long enough, most folks adopt a persona. Some are airship captains, others zombie hunters, mad scientists, or clockwork cyborgs or whatever... it really goes on forever. Mike is the Steampunk Scholar, and he's that with flair. After his panel we exchanged cards and promised to be in touch. Life being as it is, it took well over a year for us to catch up and talk. It actually started to become an in joke among the folks who knew us in the community: the Steampunk Scholar and the steampunk cultural historian kept missing one another.

In the end, all it took was the right window in both of our schedules (not easy with kids, courses, and everything else in the mix) for a phone call to Canada.

"The Steampunk Scholar": Mike Perschon (2011) (photo courtesy of Lex Machina)

I asked Mike what he thought of the growing steampunk scene and what... well, what it was. He laughed. A friendly, knowing chuckle. I stumbled around for a little more context, just having dropped the mother of cultural bombs on him (i.e. "quick... define this").

"As a cultural historian," I said, leaning back to take a sip of tea away from the microphone I'd jammed up against my propped up iPhone, "part of what interests me so much about steampunk is the fact that it's something that different people are seizing on for different reasons. There's an element of this that seems to be... well, I'm deliberately avoiding the word *zeitgeist*, but the problem is every time I say I am deliberately avoiding the word *zeitgeist*, I use it." We laughed. Mike's good people and good people laugh a lot. It makes the world a better place.

"But there's something in what's happening right now in our culture with respect to technology," I went on, "with respect to our relationship with how we define ourselves to other people, to the imagination, to what's possible, and to all of these devices and gadgets that are increasingly kind of both pervading our lives and disappearing that makes people want to grab onto steampunk. It's a catchy word, but also the little components of the aesthetic of it. People somehow create an anchor for themselves in gears and goggles."

"I think there's definitely a component of it that are doing that. I think there are just as many who are doing it because it's cool."

"That's true," I responded. "I know the dimensions of a cultural phenomenon. They're fractals, deep, complex, replicating patterns. You can lose yourself in them, or just wear it on a T-shirt."

Mike mentioned a recent essay by Jess Nevins, an important scholar in the field, that argued against the "anchor" idea I'd just mentioned. But he added, "I think you are right, though, I think there are still a lot of people who take this as serious as cancer. For them this is a lifestyle, it's an ideology, it's a way of looking at the world. And these are the ones who say there must be punk in steampunk and it's very serious for them. But some just happen to be part... just like, this looks really *cool.* They love it at the level that they love it. Like talking to Greg Broadmore, the guy from Weta Workshop who makes the ray guns. He's a pacifist. He just loves the aesthetic of guns." (Greg's awesome, by the way; my chat with him in New Zealand was a shining highlight of my journey. You can find our interview in the *Historian's Notebook* (see Prologue). There was only so much we could fit between these covers.)

Mike went on, "He likes the way they look. I mean, you could probably look deeper for an ideology, but in that case it's: 'This looks cool.' So I think that's the run of this spectrum and this is where we end up getting these pissing matches over the definitions. One person wants to define it because it's about their lifestyle, and another person just wants to say, 'No, it's just cool.'

"I don't want to be the guy who comes and says, 'Okay, this is what steampunk is' without ever having studied it. And that's what really bothered me about a number of the pieces that I read early on. You could tell they'd read *The Difference Engine* and maybe Michael Moorcock, and then they drew their conclusion about what steampunk was from those things."

I like this guy. He thinks before he speaks. In a world of one-sentence answers, it's refreshing to come across someone who understands the power of his words and lets his conclusions follow his thinking rather than the other way around.

This question of authenticity really struck a chord with me. Who judges what's "real steampunk" and what gets dismissed as hipsterism? I brought up my research on the Grateful Dead in the late 1980s and early '90s.

"There was this dynamic in the Deadhead community," I said, "of dismissing newer fans with: 'You're just a Touchhead.' Meaning they were the people who came in when the song 'Touch of Gray' became a Top 20 hit in the late '80s. And so they didn't really *count*, right? And it became this image over substance lifestyle thing, where the people who had been part of this rich Deadhead culture since the '70s felt that these new kids somehow didn't count.

"But even at the surface level, there was a little something real about it. The new kids were tapping into something with a rich past. I've always thought of that kind of cool as a gateway drug. That when something becomes really popular and starts to hit a mainstream popular consciousness, it necessarily gets diluted. It always ends up being appropriated and thrown into things."

"Yeah." Mike got it right away. He's a culture guy too. "This kind of thing is relevant for me as an academic," he continued, "and this is what I think a lot of people don't get—they're like: 'Well, Justin Bieber isn't steampunk. I can't believe you even look at that.' Well, I have to. That's the scope of the inquiry here." Brian and I couldn't agree more. We cross paths with the legendary teen crooner in Chapter 12.

"It's like language," he said, nailing it better than I could. "Language changes. And there's nothing you can do to stop that. So if you're a linguist, you're going to rigorously study this. You don't get to stand on the street corner and bitch about the fact that students are now using the word 'random' in the wrong way. The language changes. You can correct them as best as you want; say that is not admissible on a formal essay. But the language is going to change and that's the same thing, I think, that's happened with steampunk—whatever K.W. Jeter meant to say with it, which he's admitted many times now was just a joke, it is what it is today.

"The term is changing. The boundary is fuzzy. So now we look at it and we say, 'Okay, well what's common to it? What is it that people think this is?' What frustrates me is when people come in and they say, 'I'm going to tell you what steampunk is.' And they go in and really rigorously sort of try to identify at least a few things that are common to it, like some of these taxonomies where you got airships and Tesla coils... And I think, wow, that's micromanaging the whole thing."

I completely agree, of course, but I have to push back a little: "But there's an extent to which in order to talk about it in some way you've got to kind of define the edge of the sandbox, right? Of course, the harder and tighter line you draw, the less useful what you're analyzing becomes."

Mike agreed. I envisioned an accompanying nod somewhere in Alberta. "There are people who will just say, 'It's Victorian science fiction' and I reply that it's a useless term for this, because it's not Victorian science fiction. Victorian science fiction is arguably the Edisonade or Jules Verne—the people writing in that era."

"Right." He is right. Ain't no Victorians writing science fiction now. They're dead.

And then Mike nailed part of it. Or at least started to... it's Jell-O to a wall, but he's good: "Steampunk is 21st century or late 20th century people looking back on that period, and then writing about this romanticized or dystopian world—it's an idealized version, neither a negative nor really joyously positive version, but it's always hyperbolized. And that's not Victorian science fiction. Victorian science fiction was imagining *getting out* of the corset."

The Steampunk Scholar just handed me a vital piece of the puzzle. Steampunk is about putting the corset *back on* (well, comfortable corsets anyway). It takes the trappings of an age gone by and applies them forward to our own. This was part of what makes steampunk so exciting. It looks back to look forward. This idea of "crossing the streams" and setting scientific, technological fiction in the past is, culturally-speaking, a pretty new beastie. It's a creature of the second half of the 20th century, a new flavor of fantastic tale that sets technological imagination in the past rather than in the future. And this idea of a spectrum was significant, too —there are as many ways to engage with steampunk as there are steampunks.

At the con I bought goggles and a pair of gaiters. A surface touch, but they helped my vaguely anachronistic combination of old re-enactment gear—Sherlock Holmes-esque long coat, cufflinks and pocket watch—in a direction that pushed the edges just a bit more. (Yes, I really do wear those things most days. My day-to-day watch that I've kept and worn for 15 years is a Swiss Army pocket watch —I'm one of those people on whom a regular wristwatch can't keep correct time. My grandfather was the same. Something electromagnetic, or perhaps psycho-chrono-somatic...) My costume still pretty much had a "glued some gears on it," look but I was starting to step up my steampunk style, if only by a notch.

Steamcon had been a boon—a good first convention to say the least—but questions remained. Lots of them. Is steampunk a "genre"? Sure, but it's also an aesthetic, a subculture, a movement, a philosophy, and so on. But those are manifestations more than definitions.

Scholar on the Spot

> *Because I work in the time travel industry, everyone assumes I must be a scientist. Which is sort of correct. I was studying for my master's in applied science fiction.*
>
> **— CHARLES YU**
>
> *How to Live Safely In a Science Fictional Universe* (2011)

Flash forward a year to Steamcon 3, 2011. This time, I was on the other side of the podium. Byrd and I were hosting a sneak preview—an early version of the Vintage Tomorrows documentary—and I'd been asked to sit on a panel as well. As Mike and I had joked later in our chat, I'd reached the Darth Vader moment. You know: "The circle is now complete. When I left you, I was but the learner; now I am the master."

I was now "an expert." This might sound kinda self-aggrandizing, but it's just what happens to anyone who spends enough time with a given topic. I'd lay good money on the table that you're an expert on something, whether you know it or not. If you have doubts about this, check back in with yourself after you've finished this book. Every person we've spoken with started somewhere—with a simple spark of interest or a single question. Don't count yourself out.

But back to steampunk (explicitly, that is...), I couldn't come out of all this talk of "what is this thing we call 'steampunk'?" without developing some opinions. There's no neutrality on a moving train, right? I knew I agreed with Cherie about fun, but a historian needs to put context around an answer like that.

Purely practically-speaking, how the hell was I going to be able to argue with anybody in a bar if I didn't have something to throw out on the table? In these environments, a working definition is the ante that buys you into the conversation. Asking the question itself is what binds the community together. There are few things, I've found, that steampunks enjoy (and also hate) more than defining steampunk.

"Experts" need opinions. There's only so long you can go around studying something before people start asking you: "so, what do *you* think?" And for this con, I'd been asked to speak on "Humanization/Dehumanization in steampunk." Mimi, then 8, had plenty of paper and markers to draw with, so she was good. The rest of us had a darn engaging conversation on what it means to be human in an age of machines. But the question inevitably came up.

For my first few cons, I had done my best to fend off putting forth my own ideas. I had an angle as a scholar and good reason to reserve opinions—I was gathering, not disseminating them. I made no pretense to neutrality, but I admit to being a little evasive. I tapped into my classroom socratic-fu, turning questions back on their askers, and developed a pretty good repertoire of quotes and references to bounce back as "well, so-and-so says..." That'll get ya pretty far.

So, on stage at Steamcon 3, I reverted to what I've come to think of as contemporary steampunk gospel, the Catastrophone Orchestra and Arts Collective's steampunk manifesto: "What Then, Is Steampunk? Colonizing the Past So We Can Dream the Future."

> *First and foremost, Steampunk is a non-Luddite critique of technology. It rejects the ultra-hip dystopia of the cyberpunks—black rain and nihilistic posturing—while simultaneously forfeiting the 'noble savage' fantasy of the pre-technological era. It revels in the concrete machines as real, breathing, coughing, struggling, and rumbling parts of the world. They are not the airy intellectual fairies of algorithmic mathematics but the hulking manifestations of muscle and mind, the progeny of sweat, blood, tears, and delusions. The technology of Steampunk is natural; it moves, lives, ages, and even dies.*
>
> — FROM **STEAMPUNK MAGAZINE #1**
> (2007)

It doesn't get much more human than that. Certainly spoke to the theme of the panel, and it bought me some time. Fortunately, it also got me thinking. Buying time ain't no good unless you spend it.

Byrd and I had a blast screening the film. I was amazed by his wizardry, and Mimi was tickled eight separate and complementary shades of pink at seeing herself on the big screen. We took questions afterward, and I was so dazed that I still only recall one with any degree of clarity. Somebody asked: "What did you learn

SteamPunk Magazine: The First Years, Issues #1-7. Not the Steampunk Bible, but certainly among its holy texts.

doing the film that you hadn't known before?" Byrd gave one of his trademark combination charming/understated/profound replies, and all I could come up with was "I shouldn't shave." Seriously, I looked like a pudgy 20 year old. The beard is back to stay.

After a brief, but deeply inspiring handshake and chat with Jake von Slatt (if making is the religion of steampunk, Jake's at least an Archbishop), we rushed out the door—Byrd to grab his luggage, me to drop Mimi off at her mom's place. A red-eye flight later, we were in New York, repeating the process at New York Comic Con—this time with Brian, Cory Doctorow, Paul Guinan, and Anina Bennett at our sides. It was a glorious whirlwind, and just what I needed.

Travel, even of the craziest sort, clears my mind. I think it has something to do with shaking loose the patterns of everyday life. As we flew, ran, cabbed (with real-time updates on Occupy Wall Street from Cory's smartphone—man is that guy ever tuned in), I continued to ponder this many-faced beast that is steampunk. What the heck was I going to write in this book?

Well, it was a matter of putting a few pieces together and...

Asking the Right Questions

> *Do not be afraid of the past. If people tell you that it is irrevocable, do not believe them. The past, the present, and the future are but one moment. Time and space, succession and extension, are merely accidental conditions of thought. The imagination can transcend them.*
>
> **— OSCAR WILDE**
>
> *De Profundis* (1897)

I still can't entirely decipher the scribbles I made in my notebook as our cab whipped through New York City, but I'm used to this. I can't always read my handwriting. I take notes to more firmly plant the ideas in my head, and to jog my memory after. So long as the idea pops back out, I'm good.

Fortunately, this one did.

In our chat, Mike Perschon used language as a metaphor (as you may have guessed by now, I love a good metaphor, and language itself is among my favorites—did I mention that I liked that guy?), but there was more than metaphor here. From that first issue *SteamPunk Magazine*'s masthead proclaimed its mission: "Putting the Punk Back Into SteamPunk." They'd underlined something important there. Could you take "the punk" out? Aha! Steampunk is is a compound word.

The esteemed Mr. Jeter was riffing off of "cyberpunk" when he coined the term, but that doesn't mean it was born whole. Just like its cyber sibling, the word steampunk is a mashup. In each case, the meaning behind the whole became greater than the sum of its halves. There might just be some value in giving this word a little etymological what-for.

So I asked myself: "what *is* 'the punk'?" Hm. As I looked at the catchy moniker in this new light, it occurred to me that there was another question there too: "what's 'the steam'?"

You see, English has a beautiful way of absorbing and blending words. It's one of the things that makes it one of the most complex and difficult languages in the world—it's constantly evolving—hacking, splicing, and tweaking anything that touches its bowl of the linguistic soup of human communication. This is why a benevolent universe gifted us with *The Oxford English Dictionary*

So, I looked up "punk." I'll spare you the detail here (I'm well aware that there are only so many of us who deeply love dictionaries; if you're similarly-inclined, you know where to find the meat). If you're interested and find yourself without a library or $300, you can check out some of my research in my *Historian's Notebook* (see Prologue). The important thing here is that the O.E.D.'s entry on "punk" ultimately led my meandering mind to an unexpected place: Oscar Wilde's lap.

Stick with me here. This leap is actually going somewhere. It's more than just an excuse to flirt with a dead literary icon. Really. Because what I found was a glorious combination of transgressive sexuality and clever wordplay. If that ain't Oscar Wilde, I don't know what is.

It's no big stretch to imagine that Oscar Wilde would appreciate the hell out of most of what steampunks do. He tweaked the nose of convention, transgressed with aplomb, and never let a fact get in the way of the truth. But the word "steampunk" would have suggested something a bit more explicit to him. In Wilde's day, the noun "punk" meant a prostitute and as a verb the word had begun to develop some pretty specific connotations related to gay sex. Heck, the *Wilde vs. Queensbury* trial even got, er, straight, to the subject when Queensbury's lawyers played their scandalous trump card, calling out the great aesthete's interest in punking punks (Wilde was then forced to drop his libel charges when the court decided that he had indeed been "posing as a Somdomite [sic]"). In sum, he would say it better than I ever could, but it seems quite safe to speculate that Oscar Wilde would have (off the stand—homophobes held the gavels) admonished us all to punk the past with a passion.

I'm taking a few historical liberties (I have impunity—this is steampunk, after all), but I find it hard to believe that a man who appreciated witty play as much as the esteemed Mr. Wilde could do anything other than agree with Mimi about "chickens and teapots." Nor can I imagine him putting up a fight against Cherie's admonitions about fun and play. History was fair game for messin' around as far as Wilde was concerned. In his 1891 *The Critic as Artist* he said that "the one duty we owe history is to rewrite it." We're talking about playing with daddy's toys here, exercising our right (and perhaps obligation) to goof around with the stuff of the past. Rules are there to be broken, right?

In contemporary parlance, "to punk" has taken on a more PG-rated image. After all, Ashton Kutcher's "Punk'd" has 64 episodes worth of celebrities on broadcast television from 2003-2007. No, the censors weren't asleep at the wheel—"to punk" now (also) means to mess around with, or to play tricks on. The word has a lot more "family-friendly" play in it now that we're no longer (just) talking about the kind of tricks you "turn."

A Note from the Notebook: The Fun Part Is the Punk Part

What actually brought Byrd and I to Victoria was the chance to have a cup of tea with Ann and Jeff Vandermeer. Though they'd likely deny it, Ann and Jeff are in many ways the "first couple" of literary steampunk. Their anthologies are definitive and Jeff's 2011 *The Steampunk Bible* is the go-to overview on the topic (and my constant companion throughout this voyage).

Our complete interview is in the *Historian's Notebook*, but Ann's thoughts on "punk" can't be left out of this discussion:

"To me," she said, "the fun part was the punk part, because I mean steam is steam and it's not good for my hair but the *punk* part was really cool because I grew up during the time of punk rock music. I even played in a punk rock band for several years and so the punk part is what I really liked, I liked the idea of taking history and just twisting it and that's pretty much what steampunk was doing for me is reading about these fake histories, these things that didn't happen but could have happened and wouldn't have been cool if it did, and so that's what I really liked about steampunk.

"The whole punk movement was about just doing it yourself. Just diving in and doing it. A lot of the musicians that came out of that time period like the Sex Pistols and The Clash and all those guys—they didn't know how to play music, they weren't formally trained in their instruments, they just picked it up and they just went. That to me was the whole attitude of the punk movement is they made it themselves, because at that time you had bands coming out that were doing these album-oriented rock with 30-

minute guitar solos that would put you to sleep. The punk songs were two and a half minutes, boom, hard, fast, done, but it really got you right in your guts. That was to me the heart of punk—it's just grab it, make it your own, do it yourself."

Ann had just tied together a subculture history thread that had been eluding me. The punk movement does fit in with all this. Not just as a word, but as a way of looking at the world.

"I see so much of that with steampunks," she continued, "because they're grabbing these technologies. With punk rock music it was the music but with steampunk it's the technologies, they're grabbing these technologies and they're making them their own, they're making things themselves, they're taking pieces of this and pieces of that and putting them together and making something brand new."

Hence the inception of this entire project. All that we're missing is Brian and a pitcher of Pike Place IPA.

We'll talk more about punk as both noun and verb as we continue our journey. As a noun, it also has 20th century roots in the idea of an amateur or inexperienced person, which of course relates to the radical "do it yourself" philosophy of 1970s and '80s punk counterculture—who the hell are you to tell me I can't? As we'll see, steampunk is nothing if not hands-on.

So, how does one go about punking steam? Well, if there's such a thing as a "right" place to pursue this elusive question, the Victoria Steam Exposition would have to qualify.

Snark Hunting in Victoria

Victoria, BC

> *"Just the place for a Snark!" the Bellman cried, As he landed his crew with care;*
>
> **— LEWIS CARROLL**
>
> *The Hunting of the Snark* (1874)

Held at the historic Fairmont Empress hotel, the Victoria Steam Expo was a creature all its own. Loose, open-ended, informal, and intended as a culture-hack in itself, it wasn't a convention at all. It was more a gathering of similarly weird-

Attendees of the 2012 Victoria Steam Expo pose for a group photo. Amongst others, pictured are Captain Robert Brown of Abney Park, Airship Ambassador Kevin Steil, James Carrott, festival organizer Jordan Stratford, and Girl Genius co-author Kaja Foglio. (photo courtesy of Byrd McDonald)

minded people. A weekend salon, in the old cultural sense of the word. A contemporary bohemian gathering. But like any steampunk con, its cultural content ranged from the deep and cutting to the cheap and trite. Dealers, speakers, and attendees all bumped around together in a single large room beneath the Fairmont Empress's famed tea room. There was a stage, and a schedule, but nobody fetishized either—to the point where speakers sometimes wondered where and when they were really expected. In short, it was a cultural playground.

The convention's ringmaster, Jordan Stratford, kicked things off with a short speech that amounted to "You all know why you're here. You love this stuff. I'm not the authority, you're the authority. We can all do this and you take this thing, you take this, you play with it, you break it, don't be afraid of it. Talk to one another. This is your event—do what you like with it!" and then promptly declared the rest of the Inaugural Address to be an "open mike."

There was a little stunned silence, but then folks started popping out of the woodwork. "I'm a steampunk gardener and would really like to talk to anyone else interested in steampunk gardening..." and on it went. The conversation had begun, though lord help me if I had any real idea of what steampunk gardening might look like. That was a new one even to me.

These folks were punking the very idea of a convention itself. Full frontal cultural play. It was already quite clear to me that restrictive boxes need not apply. Film crew in tow (or maybe the reverse—we'd become our own circus by this point, and lines blurred readily and rapidly wherever we went), I resolved to dig a little deeper.

Byrd and I cornered Jordan later in the weekend and asked him about where this event came from. Most of the events I'd been to were modeled strongly after a traditional convention or conference. GearCon ended up being kind of informal just by accident—but this was more deliberate, and tasted more of countercultural happenings in a more bohemian vein. This was one flavor, though, not the whole dish. This contemporary take was widely and wildly inclusive.

There were folks of all ages, shapes, and sizes (not so much colors—this remains a challenge for steampunk, more on this in a couple chapters) Robert and Kristina from Abney Park had brought their kids, and kindly signed a copy of their new CD for mine. I'm pretty sure I offended Robert by likening his band to the Grateful Dead, but I meant it as a compliment—both are, in my opinion, bands that you *experience*. The music just means more once you've seen the amazing energy of a live show. But I digress.

Jordan was nothing if not outspoken and direct. He was well-rehearsed and one heck of a public speaker. A good trait in a minister-cum-artist-cum-writer.

"The Victoria Steam Exposition is an art show," he started. "It's an art happening that is different from your regular drywall airport hotel con in that we use actual vintage or period settings. We've hosted it in an 1890s castle. And we've done it in this venue before too.

"It is an interactive art experience that attracts costumers and musicians and authors," he continued. "We just put all these weirdoes in a bucket and let them riff off each other's insanity and it's art inspiring more art."

He's just articulated the reason this con was the perfect capstone to my journey through steampunk convention-land. It's a living, breathing thing. It's chaotic and alive, an inspiration to do, build, and make. I nodded as Jordan continued. This guy could talk through a hurricane.

"People come to the event with very little knowledge of steampunk and get hooked. We have a whole family that wrapped their lives around steampunk. They'd never heard of it, when they came to the first show. Now they have three generations within the family making steampunk art. This is normal for us. This kind of buzz and this sense of dynamic interaction with artists keep people coming back."

Steampunk is a Snark. Try to put it in a box and I wish you luck ever finding that box again. Any meaningful definition of steampunk has to include a wide variety of approaches—from the deadly serious to the wildly irreverent and everything betwixt and between. There are commonalities, though. Art, inspiration, questions, and play are at the heart of steampunk's challenge to our cultural status quo. Steampunk punks itself. It's a subculture that messes with subculture, hosts unconventional conventions, and finds its commonalities in difference. Chickens and teapots indeed. Well played, Miss Mimi Carrott... well played.

Digging Into the Past

With a firm start James pushes deeper into the past and the future. Following a fascinating talk with world renowned author Scott Westerfeld we end up in the audience of a particularly good discussion at Seattle's Norwescon where steampunk is defined, debated, and redefined. Next James explores the real history behind steampunk, revealing the shocking similarities between the late 19th century and our contemporary Information Age.

The Past Is About 75 Feet Deep, Depending on the Season

James H. Carrott (Seattle, WA)

MARK THOMSON (we'll spend time with him in Chapter 10) begins his brilliant *Henry Hoke's Guide to the Misguided* with a pair of quotes that has become one of my favorite statements on how we perceive and use history:

> *Very deep, very deep, is the well of the past.*
>
> **— THOMAS MANN**

> *The past? It's about 75 feet deep, depending on the season.*
>
> **— HENRY HOKE**
>
> *Henry Hoke's Guide to the Misguided* (2007)

This may, at first blush, appear to be just a folksy tweak of the nose at wise words from a literary luminary. And okay, it is. But it is *also* a pretty darn profound statement about the nature of our relationship with the past. Classic Mark Thomson, by the way, is profoundly poking fun at profundity. You see, the past isn't actually all that mysterious. We can all reach it with a little imagination and a shovel. What's more, how deep we dig really does depend upon the season.

Steampunk draws its past from a particular depth. A particular set of strata, at least. What's more, the depth steampunk digs to is highly dependent upon the season—our season. The reasons that steampunk reaches back to the 19th century —why it punks "steam" as opposed to, say, diesel or wind—have everything to do with today. Steampunk is happening *now*, and the cultural weather patterns of today have a particular resonance with those of a specific yesterday. This chapter is about those connections—the echoes and eddies that seem to be drawing our Information Age minds increasingly toward the Industrial Age.

It all starts with adventure.

All Those Futures and Pasts Bumping Into Each Other: Scott Westerfeld

As I was assembling this draft for publication, I went back over my list titled "People I Really Want To Talk To." I'd crossed out a lot of names, which always feels good. But one name at the top of that list felt like a glaring omission: Scott Westerfeld. Scott is an internationally-bestselling author, a superstar of the Young Adult genre whose most recent *Leviathan* series is a brilliant romp through an alternate history populated with living airships, walking tanks, and a very savvy young female protagonist. We'll talk about other indicators throughout the book, but one I hadn't yet examined in my writing was the turn of Young Adult bookshelves from supernatural romance (vampires and werewolves and kissing, oh my!) to steampunk.

Leviathan Trilogy

If your humble authors had been a few decades younger when Scott Westerfeld's *Leviathan* first came out in 2009, it would have easily been our favorite book of all time. Even as adults, the *Leviathan* trilogy ranks pretty high on both our lists. This stuff has serious appeal for kids of all ages (yep, even the 40 and up crowd).

As I take my final pass through this chapter, Mimi, Beatrix (9 and 6, respectively—we're time jumping) and I are about halfway through the second book—reading two chapters aloud each night. I lead, but make Mimi take over from time to time. B loves the illustrations and the "awesome girl" but doesn't like "the boy parts" as much—she wanders off to draw or just snuggles up and falls asleep. Mimi, on the other hand, has conned me into many hours of past-bedtime historical conversation, grilling me on politics, science, and technology. The other night she actually had me walk her through an illustrated (3 pages of my notebook!) history of firearms, from matchlocks to rifling, repeaters, and metal cartridges—all starting from the simple question of why you'd want to use pressurized air rather than gunpowder to shoot from a hydrogen-breathing living airship.

This is history coming to life through fantastical fiction—one of steampunk's greatest strengths. *Leviathan* knocks it out of the park. It's sure hit a home run in our little apartment.

It was painfully clear from our conversation back in Chapter 4 that Cherie Priest knows her stuff about steampunk and literature. She also knows a lot of brilliant and amazing people. So I dropped Cherie a quick email asking if she happened to know Scott. She did, of course. Within half an hour, I received an email from Scott, who said he'd heard of the Vintage Tomorrows project and would be very pleased to speak with me. I was stunned, but not so stunned that I didn't reply immediately to schedule a time to talk. We aligned schedules and set up a call.

Once I'd navigated the complexities of making an international phone call (Scott lives in Sydney, Australia when he's not in New York City), thrilled to hear a ring, I was almost taken aback when my call was answered immediately.

"This is Scott."

"Hi Scott, this is James Carrott."

"Hey, how ya doin'?"

"A little flustered. My apologies that it took 20 minutes for me to negotiate the ability to call you." I'm a pretty fastidiously prompt person, and it truly drove me nuts that I was so late. Of course this kind of thing only happens when the person you're making wait is pretty darn important.

"That's odd," he said, not in the least unkindly.

Relax and just start, James. "You've been writing science fiction for a long time. Correct me if I'm wrong, but the *Leviathan* books seem to be your first step into a deliberate and specific historical past."

"No, that's correct," he said.

"So what brought you there?" I prodded. "Why jump back instead of forward?"

"I think one of the things that happened is," he was clearly searching for the right place to begin. The place he chose surprised me. "I've always just loved steampunk for itself," he said earnestly. "I mean even before it was the sort of well-thought-out literary genre that it is now." Right. This was going to be fascinating.

"I think the first time I ran into something steampunk," he went on, "was when I went to Disney World when I was like 11 or 12 and went on the *20,000 Leagues Under the Sea* ride. It was that sort of, well, I mean to see a Victorian submarine was really cool. And there was something about that experience and about those movies of that era that brought Verne or Wells into modern movie-making. Those were always really interesting to me as a kid. I mean *20,000 Leagues Under the Sea* as a movie is a really weird thing. On the one hand it's made in an age where there are real nuclear submarines, but it's set in the 1800s and it's basically about a terrorist with a nuclear submarine. The way all those futures and pasts bumped into each other was really cool for me."

Scott was hardly the first person I'd spoken to who cited Disney's *20,000 Leagues Under the Sea* as an early influence—that film set an aesthetic tone that pervades steampunk to this day. What distinguished Scott's response was an interesting nuance. He recognized it as steampunk. Not just as a cool-looking interpretation of a Jules Verne novel, but as an example of playing with time—meshing the 1950s with the 1860s. He caught the underlying thread connecting past and future, and *that* was what inspired him.

"Another thing that happened with *Leviathan*," he continued, "was that after Uglies was such a big hit, I realized that I could sort of do anything I wanted. So I made a list of all the stuff that I liked: bioengineering, air ships, and World War I were all on that list and it just struck me that steampunk was the right genre in which to do a collage of all those things, 'cause steampunk lends itself very well to collage."

"So you clearly think of your work as steampunk?" I asked.

"Of *Leviathan*? Yeah. Well, you know, technically there are diesel engines in the book..." he admitted.

"That's a nit I don't pick," I responded honestly, "but you're sure right that some people do." As you know, I view steampunk as a pretty big sandbox.

"Right." He didn't miss a sub-genre beat. "I like the word dieselpunk if you are doing something like 'weird World War II.' I think that makes perfect sense. But to me, World War I is the dividing point where modernity goes from being optimistic to being pessimistic. Because when you put the words machine and gun together, they both change. At that point, war is no longer about a sense of adventure and chivalry and a way of testing your nation's level of manhood; it's become industrial, and horrible. So playing around with that border between optimistic steampunk and a much more pessimistic dieselpunk, which is more about Nazis, was kind of interesting to me because early in the war we were definitely kind of on the steampunk side of that."

"You've written a fair amount of post-apocalyptic stuff, as well," I said, shifting gears a bit. "Does that feel very different to you? The way that you approach the material that you're writing in the past—is it a different project for you or is it just a different side of a coin, or is there some other way that you think about it?"

Scott paused before replying thoughtfully, "Well, I think my post-apocalyptic stuff is kind of nostalgic. I mean, if you're a well-rounded science fiction reader, reading *Uglies*, you'd recognize most of the tropes: the ruined city, the outside of civilization savages, and so on. It's fairly nostalgic science fiction. Maybe not for the core readership of 12 and 13 year olds, for whom it's all new, but if you're a well-read science fiction reader, I think both *Uglies* and *Leviathan* are based on the same kind of thing in that they're both pretty nostalgic, from a literary point of view."

NOT NOSTALGIA FOR AN ERA, BUT FOR ITS FANTASIES

I wanted to know more about his ideas on nostalgia. It's a strong word that tends to taste of uncritical sentimentality. Scott's stuff is really savvy, though, so it surprised me a little that he described his work as nostalgic. I pressed him on it: "Is there a tension at all, though, with that nostalgia and some of the politics that you're writing about, in terms of the British Empire and the Ottoman Empire and the Austro-Hungarian Empire? I'm sure you've heard the criticism of steampunk that it glorifies imperialism. Do you see a place for that nostalgia that doesn't reinforce a romantic ideal of something dangerous?"

"Well," he replied, "the kind of nostalgia I'm talking about is not really nostalgia for the era, it's more nostalgia for a kind of literature about the era, which is obviously not unrelated, but is another thing. The reason why *Leviathan* works the way it does, it looks the way it does, why it includes a very formal portraiture and a lot of that stuff is that it's meant to feel and look like a boys' adventure book from 1914.

"Now, in the politics of the book and in the series itself I'm hopefully updating that. One of the books I was reading when I was doing research—very much a 'boys' own adventure' book—is from 1906, *A Trip to Mars* by Fenton Ash. One of the weird things about it is that there are absolutely no female characters in this book. There's a King of Mars, and a Prince of Mars, but there's no Queen of Mars. The only time the word 'girl' appears in the entire book is on the first page, and that's because two boys are talking about the stars, and one says to the other: 'You're a queer chap; as dreamy as any girl, I declare!' and 'When you are in this mood no stranger would ever imagine you were the same go-ahead, muscular young Christian you can prove yourself to be at other times.' It's all about policing the masculinity of characters, the characters tweaking each other about masculinity and saying: 'Look, we're not girls, there are no girls in this book.' So what I wanted to do with *Leviathan* is just to reclaim that world and put a female character in it."

He'd answered my question about gender before I'd even asked it. I found myself wondering more about his young adult audience. "It's interesting to me that when you were doing your research, you looked at all this stuff that was written for young boys, that whole dime novel, Penny Dreadful era and you're also consciously writing for young adults—just of a different era. How is that different from writing this kind of stuff for an adult audience? Are there things that you assume or don't assume about how you teach and approach history or how you think about technology or relationships?"

"Teenagers are really different from adults." Scott said, pausing to draw out another fascinating point. "Adults come to everything with a lot of preconceived ideas. For example, one of the most common things adult reviewers chafe at about

Leviathan is that it is about World War I, but it doesn't have trench warfare. About once a month I get something popping up in my Google alerts about how there's no trench warfare in *Leviathan* because if there's one thing that adults know about World War I it's that it was full of horrible trenches with rats in them."

More Than Trenches

We both knew better, of course. The First World War was actually a world war whose broad-ranging impact reached deep into the Middle East (the setting for Lawrence of Arabia) to Mexico, where the Mexican Revolution and the Zimmerman Telegram played a key role in the United States' entry into the conflict. If all you can think of is All Quiet on the Western Front, ask an Australian about Gallipoli. It's not just Mel Gibson's butt-bearing feature film leap to stardom, it was a horrific battle in Turkey that took a huge toll on troops from Australia and New Zealand.

Every single town I visited during my research in Australia had a monument inscribed with the names of local boys who had given their lives in the Great War. It's very difficult to overstate the impact World War I had "down under." In fact, Mark Thomson's Institute of Backyard Studies (the reason for the aforementioned visit—Chapter 10) is restoring a WWI memorial chapel to serve as its headquarters in the now-ghosting outback town of Terowie.

"You know," I said, once we'd recovered from our trans-Pacific bout of eye-rolling, "what we're talking about now reminds me of something you said in the afterword you wrote for *Goliath* (the third and final book in the *Leviathan* trilogy) —when you talk about the fact that the series' most important departures from history are not the fantastical technologies, but the changes in the course of the war itself."

I continued, still working to get my head around this fascinating departure from the steampunk norm. "So much of the research that I've been doing around steampunk has a lot to do with how we, as a culture, are getting our heads around the changing nature of the world around us. We tend to think first of technology but the deeper stuff that you're really grappling with in *Leviathan* isn't the technology, but how the world changes because of the way these characters interact, because of the way that the story unfolds. And that, to me, speaks to a bigger question of the effect of the past on the future and how, if we change the story of our past, maybe we can change the story of our future?"

Spoiler Alert

If you haven't yet read the *Leviathan* series, intend to (really, you should), and hate knowing how things turn out before you begin a story, skip the next four or five paragraphs. They're great, but you can come back to them later and still get the gist of what Scott and I are talking about.

"Absolutely." Scott was clearly thinking about the future, and about its connection to how we understand history. "Effectively, what the crew of the *Leviathan* does is prevent World War II. That's the thing that comes out of this. World War I ends before Germany is crushed into submission. The decimation of that generation of young men don't happen. It's not about stopping the first world war, it's about stopping the second world war, about preventing all of the really horrible stuff that's gonna happen later in the 20th century. That's the indication I wanted to give."

"In a sense what you're doing is sort of fixing the 20th century in retrospect, right?" I asked. "And the prospect of doing that is a pretty mind-blowing concept. Granted, we're talking about the realm of the imagination, but one of the things that I've been grappling with as a historian is this balance between history and imagination, because honestly, history happens in your head."

"Yeah," Scott replied. "But it's also more than that, right? I mean, in the books there's a tension going on between Alek and Deryn the whole time, where Alek had the sort of more 'great man' view of history, which is 'I am here to change x,y, and z.' But Deryn has the more modern and really more sensible view of history that it's just a bunch of stuff that happened and you can do what good you can in your context, but you can't really change everything."

That was cool enough to be a good stopping point in itself, but I had one last trick up my sleeve that always served me well when conducting radio interviews. It's a silly little thing, but you'd be surprised at the gems it turns up. This conversation was no exception to that rule.

HISTORY WEARS UNCOMFORTABLE SHOES

"Well," I said, "as we wrap up, is there anything that you'd like to add? Is there anything more you think I ought to know as I start to write about your work? Anything that I haven't asked but should have?"

"You know," Scott replied, "there is a really cool thing."

See? Works like a charm. He started to tell a great little story:

"I visited this school in Chicago—one of those schools that's really good at author visits. They had this really great woman who focuses exclusively on author visits. And when you get there the kids have all read the book and they've redecorated the school for you and all this amazing stuff. When I went there they had actually created an 8-foot-tall walker and a Zeppelin-thing made out of cardboard and a lot of the kids were all done up in Edwardian dress."

I could almost hear his smile. Scott is well-known for his deep engagement with his fans. He hosts regular fan art contests and forum chats on his blog, and is extremely responsive—no small feat for someone whose books sell by the million.

"As I was doing my presentation, I started to realize that the kids were getting kind of uncomfortable. And I got this cool chance to stop for a second and say: 'So, how do you like those Edwardian clothes?' And they were like: 'This sucks.'"

This cracked me up. I've been playing historical dress-up in one form or another for decades, and you know what? It's true. We take our lightweight cotton and synthetic clothes for granted. You don't really realize that until you've spent a number of hours in a coarse woolen suit.

Once I'd recovered from my brief fit of hysterics, Scott was free to continue: "And it struck me how really educational it is to realize that these clothes are not just a look, they're a thing you have to wear all the time. I asked them 'Can you imagine wearing that stuff all the time? Having to maintain it?' Those clothes become a huge expense and a huge part of your life and a huge part of your comfort and discomfort. They're also a big part of how you're able to talk to somebody else—to really identify who you can talk to and who you can't talk to. And these kids were all, granted in some very narrow ways, living it just for that morning. What was funny was that the girls who were dressed as Deryn were like: 'These clothes are totally awesome.' They really liked the boys' clothes. So it was really interesting to see how this kind of role play was supremely educational and supremely instructive, in a way that nobody was intentionally trying to build into the experience. The teachers just said: 'Hey, dress up like the book.' They weren't saying 'We are going to teach you about corsets now, and how uncomfortable they were.' It was just like dress like the book and they wound up learning as a side benefit that corsets suck."

"Yeah," I chuckled, "it turns out when you actually try to set your feet in the shoes of those people, that they are a little uncomfortable."

"Exactly. That was a useful lesson they'll probably really learn because they've embodied it, and with it all kinds of interesting realizations about role playing. You know, the kind of role playing that happens with steampunk, it's useful and interesting in ways you don't expect because it is role playing. You learn one thing out of playing roles that you don't learn any other way."

And really, there's not a more steampunk sentiment than that: Do it. Try it on for size. And again, here in the realm of the written word, is a punch line that says: "Wanna really learn something? Take it out of the book and put it on."

It's the steam and the punk combined. Imagined history is only a part of the picture—you have to do it, make it, and to some extent even live it.

We Have Met the Victorians and They are Us

But why does steampunk choose Victorian shoes? Why dig to this particular depth? The answer lies in the fractal patterns of historical change. History doesn't "repeat itself." Lines and spirals are equally poor representations. It's in the wider study of patterns—chaos theory and fractals—that we can find a reasonable mental image. Historical patterns ripple, echo, and reverberate. Here, steampunk provides a fascinating point of study for a historian (at least this one)—because it's uniquely conscious of the relationships between our contemporary period and another period of rapid technological and social change.

Steampunk plays upon the harmonies between the Industrial and Information Revolutions; the 19th going into the 20th century and the 20th going into the 21st. So let's take a look at why steampunks choose to play with a particular past. You know, the one that's about 75 feet deep this season.

Here's where I ask you to play along with me for a little while. There are some seriously huge problems with steampunk's use of the term "Victorian" (believe me, we will get into those soon) but I'm going to commit this sin in this chapter. Why? Well, for one thing, the darn word is confoundingly common among steampunks. For another, it's a become a kind of cultural shorthand for a set of people and ideas that are relevant to our conversation.

Yes, We're Making *Insane* Assumptions

In point of fact, the word "Victorian" refers specifically to the British Empire during the reign of Queen Victoria I (June 28, 1838 to January 22, 1901). Used culturally, it extends to values and attitudes commonly associated with that historical moment. This is where things start to get a little fast and loose. Welcome to cultural history.

Even so, the way we're going to use "Victorian" in this chapter makes very little historical sense on a number of levels. But we're talking steampunk here—"making sense" is kinda off-topic. Queen Victoria was many years in the grave when the heroes of *Leviathan* set off on their adventures, but that doesn't make Scott's brilliant trilogy any less steampunk.

With a nod to Cherie's emphatic inclusion of America in the steampunk equation, we're going to imagine here that "Victorians" includes Americans, and extend that to other western Europeans, too. That way, we can even count Jules Verne—who was (gasp!) French, and not at all a part of Britain or its Empire—among the "Victorians." Now this still leaves us with a pretty narrow group of upper-class white folks, but we're going to play along. For now.

We're working with an image. For our current purposes, it doesn't matter. Like "beatnik" or "hippie," "Victorian" has become a stereotype; a cultural bookmark that pulls up an image in our heads. Even as a stereotype, "Victorian" remains useful as a handle. Why? Because a lot of the feelings about the future that echo in steampunk culture actually come from a particular set of educated European and American people who have, for good or ill, become associated with Victoria and her time. This is about ideas with complementary flavor—two sets of feelings and images that harmonize in an interesting way, telling us something about who we are, and who we may yet become.

Making things a little more complicated, we've got to burst part of the stereotype's bubble while holding on to some of its hot air. Why? Because stereotypes overlap, and we need to clean out the part that's just plain useless. The classic stereotypical view of "the Victorians" is that they were stiff-collared (both figuratively and literally), very set in their ways, proper, and stuffy. Ask a moderately well-educated average person what they think of when you say "Victorian" and you'll likely get some variant of the above. You'll hear about bustle skirts and tea-parties, parlors and smoking jackets, and you'll maybe catch a whiff of Empire and the privileged classes (again, more soon—don't think that steampunk misses the beat on these problems). Like most stereotypes, these images contain some elements of truth, but are, in the main, bunk.

That's the part we're bursting. The part we're holding onto is something Scott called out for us earlier—steampunk's "Victorian" is not really about the actual lives

Recapturing the imagination of the past. "The White Desert! or Frank Reade, Jr.'s Trip to the Land of the Tombs" from Frank Reade: Adventures in the Age of Invention (2012) by Paul Guinan and Anina Bennett (courtesy of Paul Guinan and Anina Bennett)

of these people, but about the stuff of their fantasies. The exciting trappings of exploration, adventure, and discovery make for great inspiration, and the imagery is powerful stuff. We're looking back not so much at these actual people, but at *their image of themselves and their feelings about the cultural change that was happening around them.*

But there's more to the magnetic appeal of Victoriana than just the romance. It's about understanding our own relationship to technology in a rapidly changing age.

What makes this particularly interesting is that the place I inhabit in contemporary western consumer society isn't really all that far removed from that of our stereotyped "Victorians." Don't get me wrong, I'm a highly-educated white man living in Seattle, but I'm far from a software tycoon and I have to work hard to make ends meet. I'm speaking more generally of my seat on the broader bus full of privileged people who drive the express lanes of an imperial system and benefit, in one way or another, from its inequalities.

The fact that you are reading this book makes it not too much of a stretch for me to seat you beside me on this bus. I'm stereotyping both you and me in doing this, but, like "Victorians," it provides us with a useful handle. There are more reasons for which you could take exception to this than I can even begin to lay out here, but I'm going to ask you to set them aside and continue to play along with me for a bit longer.

Because it is no accident that imaginative anachronists have seized upon the 19th century as a focal point for their riffs on contemporary challenges. Our own age's conversation about technology can be read, from a certain point of view, as the equivalent of the Victorian dialogue on steroids.

We get a lot of the stiff-necked, old-fashioned image from our own poor assumptions about the past. The fact is that old photographs look so old because they have aged and faded over time. They were (relatively speaking) shiny and new when they were taken. Sure, starched collars are stiff but the main reason people look so formal and stuffy in 19th century photographs is that early cameras required people to stay very, very still in order to get an image that wasn't horribly blurry. Photographers literally employed metal stands to hold people's heads still.

Thanks to our friends Margaret Killjoy and Libby Bulloff, Mimi and I got to try it out. It ain't easy. We had one head brace to work with—just enough to keep an excited 9 year old in something approximating focus. Wilbur the chicken had the advantage of taxidermy. Me? I maintain that years of historical discipline allowed me to channel the deep inner stillness of past-fu. Even so, keeping completely motionlessness for 45 to 50 seconds is harder than you think.

Much like Mimi and me (but not at all like Wilbur), Victorians were far from stuffy and backward when not locked into a stiff head-brace. In fact, they tended to think of themselves as pretty wildly progressive. And they weren't far off in thinking so.

Take Sherlock Holmes, for example. Far from quaint, he was the CSI of his day, obsessed with cutting-edge technology. He was never far from a telegraph, complementing his famed deductive powers with instantaneous communication.

Father, Daughter, Chicken, Teapot (2012) (photo courtesy of Libby Bulloff and Margaret Killjoy)

He kept a chemistry lab, employed a microscope, and leveraged forensic methods still critical today, including the use of trace evidence, fingerprints, ballistics, handwriting analysis, and psychology. Imagine being able to identify an individual typewriter by the inflection of the keys—that took technological chops. Granted, phrenology (the study of the size and shape of the human skull in order to determine personality traits and intelligence), has been solidly debunked, but on the whole, Sherlock Holmes was hot. And he knew it. Recent portrayals of Holmes in film (Guy Ritchie's two movies) and TV (Steven Moffat's brilliant BBC series *Sherlock*) are pretty darn on the spot.

Technology was changing the world in the 19th century—both in hugely exciting ways for those with the means to see and use it, and also in ways that fundamentally changed the ways people thought. There's a progressive feel to Victorian culture, and elements of it that really embraced science, the idea of progress, and the future. Victorians were absolutely facing some of the very same questions that steampunk writers and artists tackle: ethics in a landscape of seemingly infinite technological capability, the role of human beings in an increasingly mechanized and abstracted world, the nature of being human in a world that feels increasingly defined by arbitrary scientific "truths."

These Victorians were also very concerned about information overload. Telegraphs allowed instantaneous communication over previously unthinkable distances. Mechanized presses and inexpensively-processed paper facilitated printing on an epic new scale. A new class of clerks and information workers were buried in piles of printed records. Readers were agog with both wonder and indignation at the fantastic and lurid ideas that spread through increasingly ubiquitous novels and magazines. The gramophone even presented the possibility of disconnecting sound and time—literally hearing a voice from beyond the grave, or listening to the same performance a second or even third time.

With railroads came new social frameworks: distance was now measured in time as oft as in miles. With synchronized schedules, clocks took on increased significance and time became tangibly less relative in day-to-day life. Gas and eventually electric lighting combined with these forces to create a terrifying new world wherein even the darkest night posed no excuse not to be working. One Victorian nobleman (the Seventh Earl of Shaftesbury, in 1888) was so flustered by these changes that he wrote in his journal: "It is a just remark that the Devil, if he travelled, would go by train."

If any of this sounds familiar, it should. The daily challenges we face keeping up with social media and email, feeling tethered to our mobile phones lest we "miss something" going on elsewhere in the world, even forcing ourselves to turn out the light and get some sleep—all of these have their Victorian parallels. Consciously turning back the clock to an age when these problems seemed newer, or at least cast in more stark relief, allows steampunk to shine a spotlight on the social changes technology has wrought.

By no means were all these changes universally bad. Most literate Victorians sure didn't think so, and neither do today's steampunks. The possibility of travel at speeds up to 70 miles an hour astounded our predecessors, and it's definitely fun to video chat with friends on the other side of the globe. We miss something, however, when we take these changes for granted. And that's a good part of what steampunk is all about—reaching back to a time both very like and very unlike our own. The Modern Age.

STEAMPUNK IS HAPPENING NOW

Make no mistake about it. Steampunk is much more about *now* than it is about *then*. What's more, our "Victorians" were the very epitome of modern. They believed it quite strongly and they were right. Steampunk is—and I hope never to use this word again in the course of this entire book (scary, bad academic things happen when one invokes such demons)—"postmodern" like nobody's business. It is a contemporary phenomenon that draws its inspiration from the stuff of modernity.

Not just that, but it's a means of coming to grips with the changes that were handed off to us like a snowball already hauling ass halfway down a mountain. Even more notably, steampunk is specifically a creature of the Information Age, uniquely able to draw on a huge amount of source material from the recent past. The first arguably steampunk book I can find is from 1947: *The 21 Balloons*, a children's book by French-American illustrator and author William Pene du Bois. It's technological science fiction set in a specific 19th century historical past.

Wait... So Now Isn't "Modern"?

The word "modern" gets thrown around pretty loosely in everyday conversation, but it has a more focused meaning to historians. I don't mean to suggest that we agree on the specifics of that meaning—far from it—just that we use the word differently. I'm making a distinction between the casual, everyday use of "modern"—a rough distinction between something recent or new and something from the remote past—and the historical sense of "modernity."

Don't worry, I'm gonna spare you the historical arguments. You didn't sign on for that kind of torture. The important part is that when *I* refer to "The Modern Age," I'm not talking about *now*. I'm talking about the couple hundred years between the Scottish Enlightenment and the end of World War II.

Roughly speaking, (I can almost hear my dissertation advisor wincing as I say this), when I differentiate between modern and contemporary here, I'm referring to the differences between Nikola Tesla and Enrico Fermi, the American Civil War and the Cold War, top hats and baseball caps.

Why does this drawing this line matter? Well, we're living in a different age now. Two World Wars, the Great Depression, the Holocaust, Hiroshima and Nagasaki—these things changed the way we thought about science and put a bit of a damper on the idea that the world was a rational place.

For the broader age that encompasses "now," I use the term "contemporary"—a quick and dirty way of saying "the age I inhabit... we don't really have a consensus label for it yet." Why doesn't "modern" mean that anymore? Blame the folks who started throwing around the word "postmodern" in the 1950s.

The Information Revolution we've been living through for the past 60-70 years has our heads spinning in much the same way that the Industrial Revolution spun those of our predecessors. It's easy for us to identify with their growing pains and we've also got a ton of stuff—letters, newspapers, magazines, books, audio recordings, films, objects, and so forth—that we can dig into to explore our sympathy.

I could barrage you with a host of examples from an extensive multimedia steampunk bibliography, but that's not what this book is meant to do. Besides, the sentiments I'm talking about are summed up quite brilliantly in a single place (if you're interested, there's a reading list in the *Historian's Notebook* (see Prologue). In Paul Guinan and Anina Bennett's now iconic book *Boilerplate* (you'll remember Paul and Anina from our dinner in Chapter 5), Lily Campion writes a letter to a friend about on the unveiling of Boilerplate, her brother Archibald's "mechanical marvel," at the World's Columbian Exposition, on 27 November 1893. In it, she says:

> *We have discovered a new irony of the modern age: In a place where every thing is a wonderment, nothing is a wonderment.*
>
> **— LILY CAMPION**
>
> from Paul Guinan and Anina Bennett's *Boilerplate* (2009)

Boilerplate on the Midway, 1893 World's Columbian Exposition—a "mechanical marvel" lost in a sea of wonderments. (image courtesy of Paul Guinan)

Pause for a moment and look around you. Check your pockets and glance at your desk or nightstand. Chances are (I know—assumptions like nuts, but I'm making a point here), you'll find one or more devices that can connect to a wealth of information unparalleled in human history. And yet I expect that, like me, you kind of take it for granted. We've got so much technology in our lives that the miraculous has become mundane. Well, the Victorians felt the same way. Anina's brilliant epistolary narrative (using letters to tell a story; I just really wanted to use that term) would ring as true in 1893 as it does now.

Steampunk's roots in the Victorian imagination affords a unique perspective—the insightful "what if?" juxtaposition of the Industrial Revolution and the Information Age. The parallels are powerful in simple comparison—crossing the streams amplifies this potential. Two written examples: both *The Difference Engine* and Thomas Pynchon's *Against the Day* provide powerful depictions of class struggle amplified by information technology. One visual example: Molly Michelle Friedrich's "Rise Up O Workers of the Air" (2009) takes 19th century Marxist uprising to the skies, giving an imaginative technological twist to the rhetoric of revolution.

When we compare, contrast, and combine the "Victorians" and ourselves, we highlight commonalities as well as the differences. Playing with this particular history reminds us of the humanity in the machine, and can give us a toolkit to tinker with today. And what does conscious, thoughtful, good-natured, and fun tinkering with today get us? A better tomorrow.

I won't sugarcoat the romanticizing that still goes on in a lot of steampunk, but as the various elements of steampunk community continue the process of coalescing into a movement—as makers, writers, costumers, performers, and fans spend time together at conventions, performances, and other gatherings—a political consciousness is coming together as well.

Rise Up O Workers of the Air (2009) (image courtesy of Molly Michelle Friedrich)

Time Flies, But Not Very Well

> *Don't get locked in to your own timeline, don't commit to any particular path, don't be where you are.*
>
> — **CHARLES YU**
>
> *How to Live Safely in a Science Fictional Universe* (2011)

We'll come back to the "Victorian problem" more in the next chapter, so bear with me as I jump around in time again to head back to Norwescon for more digging on just what this whole steampunk thing really is, and why it is thriving in the now. The highlight of that year's Norwescon for me was a little unexpected. I'd sure enjoyed the heck out of Paul Guinan and Anina Bennett's talk on their latest book, *Frank Reade: Adventures in the Age of Invention* (2012), but I confess to having become a little skeptical about what more I could learn from a general panel on "steampunk culture." I couldn't have been more wrong. After all, Paul Guinan and Anina Bennett along with Claire Hummel and Diana Vick couldn't help but be entertaining conversation.

Anina was the panel's moderator. As you already know from dinner (Chapter 5) and the quote on "wonderment" above, she's one hella smart lady. I knew that from the first time I sat down with *Boilerplate* and it's fabricated documents that rang more true than their "real" archived brethren. Seriously. Paul's artwork is brilliant, defining the book's insightful, yet playful character at a glance, but don't forget the words. I always expect smart things from her. What I didn't expect was that she'd turn this panel back around onto the audience in a way that spoke directly to my research. I honestly couldn't have asked for better.

Once the panelists had gone through their introductory banter, Anina began: "So, we've been throwing around the term "steampunk" a lot—it's in the title of the panel, obviously. I just want to ask the audience, is there anyone in the audience who is kind of wondering what steampunk is, or do you all pretty much know what steampunk is." With a nod in my direction, she smirked, "You, James..."

"*I'm* wondering what steampunk is!" I responded, evoking a round of laughs from my friends on the panel and in the audience.

Ever on her feet, Anina introduced the topic briefly and opened the floor. Panels at these kinds of cons turn easily into fun and informative conversations given the right moderator (it takes poise and finesse to manage the inevitable enthusiast who combines earnest loquaciousness with a notable dearth of social skills) and Anina is the right moderator. Her first question (bless you Anina) amounted to: "Why *now*?"

"Generally," she said, "I would say that steampunk is a subculture that grew out of a literary genre. It started quite some time ago as a literary genre, and the steampunk label actually was a play on the cyberpunk science fiction subgenre label. And over the years it sort of started intersecting with maker culture and cosplay and all these other things, and it became a rapidly growing subculture that has in the past 5-10 years grown exponentially. How many of you came to steampunk within the last 5 years, would you say?"

A good proportion of the room raised their hands. The first guy who spoke up made it clear that this panel was going to be a gold mine.

"I think steampunk is popular because it's more fun than goth or cyberpunk. You don't have to be drenched in ennui, or be so wrapped up in despair or in dystopia."

Diana jumped in: "Okay, sure, but why now versus back in the '80s when it was first coined? What caused the new explosion of it?" I love it when people ask my questions for me.

"Because reality is filling the need for dystopia," the guy replied.

Once the laughter had died down a bit, Diana followed up. "I agree with you to a point. I think the reason it's so big now is partly because of the internet. The steampunked computers by Datamancer and Jake von Slatt and other makers have been very interesting to people. It made people realize that things that they own, the slick, soulless mass-produced things are not making them happy. We do these things with our own hands. We create things..."

A crash disrupted the conversation. Something had fallen off the room's back wall.

"Oh, it was the clock," a woman said turning back to the group.

"Oh, God," someone groaned. More laughter.

"This panel will be eternal," Anina smiled. "It will never end." The room had entirely devolved into giggles.

"Time flies," Paul added. "But not very well."

Anina kept up her brilliant moderation. "I want to hear from the audience. What are your ideas? Why has steampunk really taken off?"

"A lot of the movies and television shows have incorporated steampunk," one person answered. "TV shows like *Warehouse 13* and the new Sherlock Holmes movies."

Anina nodded, "I see that as part of the explosion."

Now the entire audience has their hands up and wants to talk. This is the heart of steampunk: good, smart, fun people with great senses of humor.

Another audience member: "It's lost technology and lost opportunity. There's an explosion of change right now. People want to look back and go to their roots. People yearn for hand-built craftsmanship."

"Are people yearning for control?" Diana asked. "Because now we're in a technology-based system where we can't control or even understand 90%..."

"Yeah, and not just technology," Anina added. "People feel like their lives are out of their control. We even live under a political system that we may or may not agree with, but we feel like we're powerless to affect it."

Now we're on to something bigger. Our technology is a part of our culture, and ultimately, it's our changing culture we're struggling with.

Another guy in the audience: "I want to go down to the hardware store and buy something off the shelf and use it to create something. I don't want to say, 'Gee, I ran some awesome code, guys, look at this. What d'ya think?'"

More laughter. Humor is important in this culture (we'll come back to this later, too). When folks are laughing and thinking at the same time, good things happen.

He went on, "I want to come up with something 3 dimensional. Maybe it's got gears and cogs and it moves. That's the joy of it, I can actually grab this thing and display it."

Claire jumped back in. "That goes hand-in-hand with the internet and the explosion of things like Maker Faire that have popped up. Everyone is willing to teach everyone else. They were willing to do it before but now there's just a venue for it."

Next to speak up was a young woman dressed as Pinky Pie, a *My Little Pony: Friendship is Magic* character (one of the hottest new shows on the geek scene. Seriously, do you want to talk fascinating subcultures? One word: Bronies.). She nervously chimed in: "Well, one thing that I've seen as being a younger person is a lot of my friends from the anime convention scene are flocking to steampunk because it's something that we can make ourselves. And a lot of my friends who are around my age or younger—cause I'm 23, so I'm still relatively young—my friends are not going to come to a scifi convention because a lot of them have the bad connotations of the dungeon dwellers."

The room erupted in more good-natured laughter.

Pinky Pie paused, smiled and continued, "And I'm like, it's not that bad... it's fun. They like steampunk because it's something that we can do, that we can colorize, we can make our own characters. It's completely our own, unlike an anime convention, where you dress up as somebody else."

"Yeah, in steampunk it's generally expected that you'll make your own character," Claire said.

"Yeah, they love that," Pinky Pie finished up. "We can do it relatively cheaply and be ourselves and have this good venue to do it."

"Thank you," Anina said. "That's a great perspective."

"Yes?" Anina pointed at a man in the audience with his hand up. "Sir?"

"I've written a poem," he said with a sheepish grin.

"I'm sorry, is this related to the topic?" Anina asked.

"It's a work in progress, but... yes." His smile broadened.

"Wow! How could we say no?" Diana added.

The gent straightened himself and started:

> *"A future more than 100 years gone has quite recently been found.*
>
> *With steam and gears and a fashion sense from Goths who have stumbled on brown.*
>
> *With a tip of the hat, to Wells' and Verne's with a hat that's well worth the tipping,*
>
> *We'll read new stories set in olden times with adventures and bodices ripping.*
>
> *We'll sail the skies in a bag full of gas to some distant mysterious isles*
>
> *To triumph for* Science! *for country and queen, with panache and mischievous smiles."*

First the room exploded with laughter, which was followed quickly by applause. The poet blushed and sat down.

Early-onset Steampunk

Victoria, BC

To cap off our inquiry into just how these two time periods tie together and why it's happening now, we're going to jump back to the Victoria Steam Expo, which we visited in Chapter 6. Before the conference kicked off, I met with organizer Jordan Stratford at The Fairmont Empress Hotel in Victoria, British Columbia—it was in many ways the perfect location for some steampunk meta-musing. The place is steeped in turn of the century splendor, having maintained and cultivated a studied elegance of its founding era. It is both Victorian and not—having opened in 1908, the hotel never got to play host to the empress for which its city was named. That said, construction began during the great monarch's reign and its furnishing and opening were early Edwardian.

Sitting in the Empress dining room I sipped tea and awaited my eggs Benedict. Rich wood columns and ceiling beams, capped with Corinthian flair and *fleurs de lis*, patterns were everywhere in deep, rich colors. While it certainly seemed that some contemporary effort at color-matching had been made, the feel is more full, more textured, more rich than the more minimalist and streamlined design of our modern era.

The artwork was period too, but not quite right. A true Victorian dining room wouldn't be caught dead displaying even Impressionist levels of abstraction. But the result was a collection of inherited grace, overlaid, remodeled, and enhanced with a layer of a contemporary conception of the past. It was a tailored history that told me more about who we are than who we were. "God Save the Queen" played on muzak.

After Jordan arrived and we exchanged some polite small talk, I dove in: "Why now? Why has steampunk become so popular these past 5 or 10 years?"

"I have a couple of theories and they all really relate to how we react to media and material culture," Jordan started. "We live in a day and age where taking a screw driver to your TV set is a heretical act. Steampunk questions this dialectic of consumption. Steampunk says that my job is not just to consume. We want to engage, make our own, break it, risk it, reinterpret it, turn it into something else, then save the broken bits for later."

Jordan went on, getting to the essence of his point, "There is a yearning for disintermediated experience. There's a real hunger for an authenticity. People want to get into things that will have history, that will have stories that are unique and personal. Steampunk is all about that. You can do it with the clothes, or you can do it with the ideas in your head. It allows us to explore our ideas about race and gender, history and colonialism. It's the idea of hacking and breaking it and using some kind of anachronistic time bending to extract the kind of lessons we need to be asking ourselves in our culture."

Not just steampunk, but culture itself "ships broken." I doff my hat to Cory, but want to take it a step further. What Jordan is suggesting here is that steampunk is about *hacking culture*. Not just material culture, but ideas as well. This is a guy who sees what's going on in our culture right now, and how steampunk fits into it.

"So," he continued, further demonstrating that (as I've always believed) brevity is far from inextricably linked to the soul of wit: "I see steampunk as a series of questions or ways of framing questions. And why this is happening now is that we live in an age of easy answers and we are complex creatures and I think that we know instinctively how palpable and appetizing those easy answers are. We can tell that simple answers to complex questions are rarely the right ones. Steampunk makes everything more ornate, more complicated, more challenging, and in many ways more difficult, and there's something adventurous and imaginative about that, about unpacking all of that. That's very powerful. People connect to that experience. A certain kind of person connects with that experience."

I thought of questions and Snarks, chickens and teapots, smiling to myself as I asked: "What kind of person?"

"You know, I think that steampunk is probably gonna wind up in the DSM6 as some kind of disorder, and it's gonna be an early-onset disorder. By the time you're six or seven, if you were the kid who took the light bulb out of your lamp, wet your finger, and stuck it in just to see what would happen... that kid's gonna grow up to be a steampunk, if they survive. Steampunks were the kids who took their alarm clocks apart and played with trains and then glued Godzilla to the trains. This never accepting that the toy or the experience or the story was the end, but rather the beginning, as an opportunity to jam, mix, and create a whole dialectic about object and idea."

The Victoria Steam Expo was its own amazing jam—so many fantastic conversations (too many to detail here), and a creative, intimate vibe. We wrapped things up on Sunday afternoon. Tired and overstimulated, I grumped my way through the ferry ride back, accompanied by a very tolerant friend who'd come along for the ride and the incomparable Kevin Steil, the Airship Ambassador, whose even-keeled kindness further underlined something I'd learned throughout my travels:

steampunks are just plain nice people (even among these, Kevin's a paragon). Kevin went so far as to drop me off at home, taking a detour on his way back to his home in Seattle's east-side suburbs. I boiled up a cup of herbal tea and sat back on the couch to ponder what it all meant.

I thought about all those futures and pasts bumping into each other, as Scott Westerfeld had said. And I kept bumping up against the dark(er) side of steampunk. The romance and costumes and imaginative gizmos are all fine and good, but they don't occur in a vacuum. I realized it was time to grab steampunk by its ornate lapels and dive in to "the Victorian Problem."

History Has Sharp Edges

This is where things get serious. History is not some quiet dead beast that will just lie down and let you play around with it. History is alive and it has sharp edges. When we play around with history—when we punk the past—James reminds us about one of the very first lessons we ever learned in school. No, he doesn't mean: "Don't flush your towel down the toilet" (that was just him... at least he hopes so). What he means is: *"Don't run with scissors."*

The past is full of atrocities and inequity. It is also filled with inventions and inspiration. This is serious stuff. What can steampunk do to change the world? Is it for good or for ill? James doesn't shy away from the darker side of steampunk, digging into "the Victorian problem."

We are deep into the *Vintage Tomorrows* journey. James, Byrd, and the crew talk with artists and activists, doubters and optimists, each with a different perspective on what steampunk can, and perhaps *should* do.

Let the controversy begin! This is the *steam* part of steampunk, tapping into raw potential and powerful historical forces. Let's start with a conversation that challenged all of our assumptions.

Man in front of the gear—China Miéville on stage at the Key West Literary Seminar (photo courtesy of Byrd McDonald)

Having a Higher Bar: China Miéville

> *"The men call it Mightblade. Not mighty." He spoke slowly. "It might; it might not. Might not meaning potency, but potentiality. It is a bastardization of its true name. There was a time there were many weapons like this," he said. "Now, it is, I think, the only one left. "It is a Possible Sword."*
>
> — **CHINA MIÉVILLE**
>
> *The Scar* (2002)

China Miéville has serious presence. His shaved head, piercings, and tattoos lend a certain air of "badass" that blends well with the mystique of being an acknowledged master of the weird and surreal. Read up on the guy and you're like to become even more intimidated—in addition to his overwhelming chops as a writer of fiction, he's got a Ph.D. from the London School of Economics and Political Science (commonly referred to as the LSE and home to *The Economist* magazine), has run for Parliament, and published on Marxism and international law. I did a term abroad at the LSE back in 1992 and (though I admit to spending more time in the student pub and at the British Museum than I did in class) I can safely say that you don't emerge with an advanced degree from that place without being a serious intellectual butt (or should I say bum?)-kicker.

My first interaction with China at the Key West Literary Seminar was directing him to the restroom before the keynote on Thursday night. My notes in my journal still crack me up: "Got to talk to China Miéville for a second today. Good thing I knew where the bathroom was." I'll admit it, I was intimidated. I had even more on the line, though, since it was China's introduction that had brought me to Key West. More on that adventure in our next chapter.

So here I am, preparing to interview one of my favorite authors and he's brilliant and badass and I have something to prove... Yeah, I'm nervous.

It's funny, though, I needn't have been. He's actually an astoundingly sweet guy.

I had just finished reading *UnLunDun* aloud to Mimi and B (the shortest of Beatrix's nicknames; she was named after Beatrix Potter, another brilliant light of early 20th century womanhood). *UnLunDun* is China's only children's book to date and it's simply amazing. The best way I've come up with of describing it thus far is that it's sort of *The Phantom Tollbooth* meets *Neverwhere*, but with China's signature surreal edge.

I've found that there are two major indicators that something has really reached my kids: 1) It shows up in their play. Example: I recently introduced Mimi to *Doctor Who* (the classic series—the new stuff is a little too scary for her yet) and she immediately built a Lego TARDIS, complete with Doctor, K-9, and Romana. 2) It shows up in their notebooks. Mimi started drawing characters from *UnLunDun* almost immediately.

When I went to meet with China, I was under strictest instructions to deliver a picture to him from Mimi. He received it with a wide-eyed grace that I wouldn't have initially expected. Truly moved by my kids' love of his book, he signed their copy (I lean increasingly toward ebooks, I already have a 16'x8' wall of books in the limited space of my studio flat, but this one I got in hardcover—it called out for tangibility) really sweetly. Once again, my daughters contributed something important to the project—giving dad an intro that calmed his nerves.

China and I sat down between seminar events in a nearby restaurant. The place was closed, but the folks running it had kindly offered to allow us to use it for interviews. Very casual place, Key West. Roosters in the street casual. Yes, again with the chickens.

Even in the most casual of contexts, there's no point mincing words with China Miéville. You cut straight to the point, because if you don't he will. It's just a simple factor of the guy being so damned smart and caring so damned much.

The girls were thrilled.

"Your work," I began, "is often put at the top of the list of steampunk 'must-reads', but it's far from traditional steampunk and I've also heard plenty of argument that you don't write steampunk at all. Do you think of, say, *Perdido Street Station* or *The Scar* as steampunk books?"

"Okay," he responded, taking stock of the initial avenue of conversation. "I should say that I think it's inappropriate to get sniffy about people characterizing your stuff. If people want to call my stuff steampunk, that's fine. I didn't set out to write a steampunk novel, but at the same time I understand why people class it that way. It's really no skin off my nose. I also think it's important because to the extent that these movements and moments and so on have any kind of reality, any kind of heuristic usefulness it is at least as much in the culture of reception, as in the culture of production. Which is to say '10,000 steampunks can't be wrong.' I mean, of course they can, but if they tell you this is a steampunk novel, I don't give a shit what you thought you were doing, clearly you are part of this thing. Or at least the text is. That can be a problem in some ways but I think a lot of the arguments about taxonomy are slightly pointless."

We've heard a lot from folks who just sort of happened into steampunk (whether they embraced it or not), but what China just said is a step deeper. He's saying that people on the receiving end of culture have a real voice in what it is. If you've studied contemporary literature, this idea probably isn't new to you, but China had just reminded me that culture is a conversation. "Is this steampunk or not?" is really only a tiny subset of some much bigger questions: "What does this mean to you? Does it affect the way you understand the world? Did reading/seeing/experiencing this *change you*?" And even more profoundly: "Did *you* change *it*?"

While all this was churning around in my head, China elaborated on his insight: "I don't want to get involved in a kind of endless head-on pin debate about 'is this?' or 'is this not?' For my own stuff I know some people consider it to be steampunk, some don't, I really don't mind. I am very interested in steampunk as a phenomenon. My own place within it, I am less interested in."

This was going to be one heck of an interview. China digs not only to the root of the question, but into the soil in which it is planted, and gives you an answer that lays out the whole ecosystem. Don't get me wrong, he's perfectly capable of pithy wit, but he rapidly (and correctly) read my interest as deep and intellectual, and responded in kind.

"Cool," I replied. So much for underlining my deep, intellectual approach. Not to worry, China gets it. "I'm less interested in labels than about the deeper questions of reaching backward to look forward and the kinds of cultural source material that you pull from. Krakens are showing up everywhere, you know. The bits and pieces, the living cities, the entities you've imagined have resonance. *The Scar* is now among my favorite books ever. I just love the image of The Armada: that ancient floating pirate city constructed of linked ships and all the cultures and social strata that give it life. I'm also really taken with the idea of the Might Blade—a weapon of potential rather than power."

"Well that's lovely to hear you say." China accepts compliments with modest grace. He's not self-deprecating, just polite and respectful. In my experience, this is a quality common among the very best thinkers. They focus on ideas, not egos. "The whole quantum possibilities thing is an old science fiction riff. I liked the idea of weaponizing that."

"What makes the Might Blade so interesting to me is that it's physical," I said. "It's an object."

I went on, trying to communicate the particular wonder I found in what China seemed to see as a somewhat simple homage. "One of the ideas that has come up in my research is that making has become, in some ways, the religion of steampunk. Not everybody is a maker, but those who are get kind of revered. I think of it as kind of like an Anglicanism. Some steampunks are fervent and evangelic all about it. Many just nod to the makers as a kind of priestly class, maybe attend church, hit the coffee hour or whatever, and go about their business. But they seem to share a kind of reverence for the idea of creating as opposed to consuming. The gritty sense of physicality that comes across in a lot of your work strikes me as really relevant to this idea."

"Well there's a paradox here though isn't there?" China doesn't miss a beat. "I'm very interested in what you say about the making culture, and it's quite elegant. It's nice to me that you're talking about the sense of physicality in my writing, but in reality making culture, building things, doesn't interest me vastly. I'm delighted people are doing these amazing artistic things, but I have no particular desire to do it myself. The question of relationships to the physicality of objects in my stuff comes, when it's conscious at least—which is a huge caveat—out of a very different source which is essentially a surrealist tradition, which is the tradition of surrealism through the magical objects of people like André Breton, Man Ray, Méret Oppenheim, and so on about the re-enchantment of the object. That is obviously related to what you're saying."

I was reminded of just how closely tied together all these ideas are—from the very start of my adventure. Surrealism is a late modern phenomenon, born in the 1920s—a kind of successor to Dadism. (You may remember Dada from our discussion of paradigm shifts in Chapter 2.) Well, the surrealists actually had an idea about what should replace the paradigm they intended to shift. It's hard to distill, but surrealism is basically about questioning rationality and supplanting it with imagination. Conscious thought had brought us world war, so why not give the sub-conscious a chance? As you might imagine, Allen Ginsberg loved the stuff. When he and fellow Beat poet Gregory Corso met famed surrealist artist Marcel Duchamp, Ginsberg kissed his feet and Corso cut off his tie. The recursive, fractal patterns of counterculture can knock even the best of us for a loop every once in a while.

China, of course, remained unphased. "The problem is that both with surrealism and with steampunk, the re-enchantment of the object is kind of Janus-faced. Because to one extent it is obviously a kind of very creative moment of aesthetic rebellion against the kind of corporatized present in which commodities are utterly anonymized. There's a whole tradition of thinking about surrealism as a response to the evaporation of the trace of the human within the commodity. So on the one hand that's kind of a human utopia. On the other hand, as with anything, surrealism and steampunk and everything else you can think of, it becomes very quick and easy for that to become commodified in turn."

"Absolutely." Dead on. Steampunk is going there, just as the hippies did.

"My own attitude to this," he continued, "is relatively, I wouldn't say relaxed—I think the corporatization of everything is horrendous—but it is kind of philosophical in the sense that I know this happens with every interesting artistic movement. I am old enough now to have seen people who were into drum and bass, people who were into glitch, people who were into grime, people who were into surrealism, steampunk, all saying, 'This is different. This is ours. This is keeping it real.' Well, yeah, give it a year. Now that's not its fault..."

"Well it's how cultural change happens," I chimed in. "Things get appropriated; they affect the culture in some way or another, and the people who are at the heart of trying to make that change move onto the next key sharp idea."

"Exactly, which is why I think it's important at an aesthetic level to try to be a little bit light on your feet and to not fall into the trap of becoming a cliché of yourself." He chuckled. "It's interesting you mentioned Krakens earlier, I love and have always loved tentacular creatures, but I think I'm not allowed to write about them anymore for awhile. Not *right now* at least. Zombies, vampires, steampunk and so on. The bar gets higher and higher if what you're interested in is something really interesting, simply because there is so much of it."

"I think there is an inevitable banalization of this stuff," he continued. "To me what is interesting about any kind of cultural moment tends to be its written culture in particular. Now in the case of steampunk it is kind of indivisible—I understand that—but to me the key questions are political questions, as you said. You're thinking about it very much as in relation to the object, which is fascinating, and not the way I would normally consider it. To me, I come into considering it in terms of its relation to relations of power. It seems to me that steampunk and its exponential rise is absolutely related to relations of power in the last 12 years."

"Well," I said, trying to draw out what I considered to be a significant political angle, "I don't think it's a coincidence that the big spike in steampunk activity happened around 2007 and 2008, just when the iPhone, the Kindle, and this huge explosion of digital content swept into our culture. All of a sudden there is this response that drives back to physical things and their roots in the past. But you're right, of course. It spins back around again and it becomes another way of selling."

"To me, though," China responded, "the big question is: what's happened? Because you've traced the cultural bobbins, you've seen all the – you know the books and all that, it is clearly not true that—which is why I think that thinking about something like Verne and Wells as steampunk is not helpful, except to the extent that they then become refolded back in because when they were writing, they were just punk kids, well, Wells was—but it's also equally not true that there's never been any retro-futures based on the Victorians and there's actually, although there's been this exponential rise in the last 5 years, there's been healthfully self-conscious steampunk for a good 8 years before that and a fair fucking sub-tradition of Victorian-influenced retro-futurism for probably 15-20 years before that."

"Well in France, certainly in graphic novels," I said, "and steampunk—by name —has been alive and well in role playing games for decades."

"Exactly," he agreed. "But at what point is there a qualitative shift and it becomes a mass culture thing? I think those questions are always political questions. This is why to me when you're talking about the Kindle and iPad and so on I think that's perfectly true, but I would see the driver as being a much more political thing. And to me, the cultural moment in which steampunk has to be understood as a mass moment, is the epoch of neo-liberalism—not just neo-liberalism but neo-conservatism as well. This is not a new thing, but the whole relation to Empire in steampunk is really, I think, a key question. And that's why I'm sort of diagnosing steampunk. For me, the key questions are really political."

IT'S ABOUT POWER

"I have kind of a tentative genealogical theory of why steampunk, why now, why then," he said.

"So please what is it?"

"It's nothing particularly original. I'm sure you've read Charlie Stross's article that caused the whole kerfuffle, a couple years ago?"

Cherie Priest brought this up back in Chapter 4. "I'm familiar with it," I said. I can see why China would call this out.

"It was a frustrating article in some ways," China said, "because there were elements I thought were absolutely right and elements I thought were not right. But I'm really glad he wrote it because it started a debate. I couldn't care less whether the science worked or not. But what he also talked about—and you see the problem is because this is a very geeky scene and as you know if you're part of any kind of geeky scene, the ludicrous overinvestment people have. I'm speaking from the inside; I'm not being mean."

"This completely rhymes with my life," I said with a chuckle. "Go ahead."

"Okay," China continued, secure in a space that was both critical and geek-friendly, "there was this kind of rage and defensiveness of their own preferred cultural product. I was discussing Charlie's article with someone and they were really angry, like he had attacked them in the street, he was saying it was bullying. I said, 'It wasn't bullying. It was someone's opinion. If you don't agree who gives a shit?'"

This is the power of the online geek kerfuffle. This reaction tells us a lot about the intensity of the steampunk community at the time. To China the argument over Stross's article was strong enough to warrant a brief tangent away from what he saw as the larger political question.

"But what Charlie raised which I thought was absolutely key, was the question of steampunk's relationship to empire. There will always be exceptions, of course there are in any cultural argument, but I think we would be kidding ourselves if we denied that the vanguard text, and I mean text in all media including visual and creative text of steampunk in the last 10 years have written out empire. They haven't written out Victorian war, quite the opposite, but Victorian Britain, there is no such thing as Victorian Britain without the fucking Raj for example, this is obvious. But in steampunk you don't have that, it is just not there. Now this is beginning to change a little bit and that's great, but essentially the question for me, is what happens at the cultural moment when there suddenly becomes a fascination, a kind of cultural investment with the physical things of a particular moment that is stripped from the political and economic situation that allowed them, that gave rise to them?"

This is a really good question, and it bears a little underlining. China's asking us how we can separate the gorgeous gadgetry from the exploitation that allowed their creation. In Chapter 3, Cory called out the fact that we don't talk about the blood in our technology. China's reminding us that the same applies to steampunk itself.

"The shorthand version of my discomfort would be I do not think you can see as coincidental the precise temporal overlap between the rise of steampunk in the last call it 12 years, maybe 15, and the rehabilitation of the empire in Britain and the U.S. This has happened very specifically at an ideological level. Since the late '90s, certainly in Britain, certainly in Britain, and I think to some extent in the States, there has been a kind of propaganda of the rehabilitation of empire. In Britain there has been a lot of talk about how, for example, we have to stop apologizing for the empire: 'we always talk about the bad things, but what about the good things the empire brought us' and so on. This is very explicit in Britain and it crosses both right and supposed left. Like the New Labour were very big on a kind of 'Britain has to stop apologizing for its past.' My immediate response is 'when did you apologize, did I miss something?'"

"It is not as explicit here," I said, not denying his line of reasoning.

Note

Since many of you reading this book are Americans, this may call for a little explanation. The United States tends to have a problem admitting to empire, though we've been working hard on ours at least since the Mexican-American War. It's the same kind of self-deception we practice when we pretend we're all "middle class." There are words you're just not supposed to use in America. We don't invade and exploit, we "bring freedom" and "open markets." Even in this economic downturn, the Horatio Alger myth still holds some sway—this idea that becoming wealthy and powerful in America is just a factor of having the right amount of pluck and a "can do attitude." It's all in what you don't say and how you don't say it.

"I don't want to be too narrow about this," China is very considered in his expression—he's speaking generally, and knows it. "I don't want to be too lame. I understand that these are not the things that are foremost in the minds of a lot of the people doing it and that's fine, but just like with any artistic movement, I think it is perfectly legitimate to do sort of cultural diagnosis and I simply think the kind of efflorescence of a depiction of a splendid nostalgic moment in which the kind of engine of exploitation allowed is not a coincidence. I think in a sense steampunk needs, if it wants to be a serious and interesting and engaged and sort of toothy aesthetic movement, it needs for example, the Belgian Congo. Well, particularly in the case of the Victorian Era you need the imperial subject. I think there has been a big debate about this and you've got various people who are really beginning to engage to this and I think that's really exciting.

"This," he says, talking about empire and mass culture, "is probably not unrelated to the question of making that you talk about because maybe one could make an argument that steampunk is too fetishistic about objects. The paradox being that folks think so narrowly on kind of the object, the making of the object and the sort of the artisanal nature of it and so on. Clearly, actually anything that is *made* has a huge kind of political, economic basis behind it. So it's no surprise that in a sense that steampunk is about abstracting out all that stuff and just looking at it as kind of incredibly sort of aestheticized and abstracted way. None of this means that this stuff can't be done interestingly or critically, but it does mean that one needs to be kind of sharp about it. Once the culture industry really gets a hold of it, it becomes to a certain extent sort of self-sustaining because as long it's cool, people can make a buck on it.

"I think," China continued, "that steampunk is an aesthetic moment that can be diagnosed culturally like any other aesthetic moment and understood in terms of the political economy out of which it emerged. In terms of what it can *teach* us, I am pretty skeptical. I think it probably can't teach us very much, beyond how if we analyze it we can work out some of the elements that got it to grow up in the first place. I am fairly skeptical about it being able to teach us anything very useful about technology or about where we go from here as well. I don't think that's its job and I'm not convinced that it will do it."

"So what *is* its job?" The natural next question.

"I don't think it has a job. I don't think it has a job any more than surrealism or cyberpunk or any other such thing has a job. It isn't a thing with a task, it's an aesthetic moment. To the extent it has a job, its job is to entertain people. If it does that, terrific. But it's something to be understood, not something to understand."

Wow. China draws a really incisive line between cultural production and change. I don't think he's necessarily wrong—I just think that we view cultural production in different ways. In a sense, the project I embarked upon from my earliest studies of counterculture has been working out the cultural elements that manifested in a given rebellion. But I see these as things that can be applied forward. I'm not sure China completely disagrees with me, but he sees a much sharper line between cultural movements and real political change. We certainly both concur that the line needs to be crossed in order to make a real difference. But what line? And what constitutes crossing it? What is counterculture, anyway? Well, we may be in different camps there, but he raised some very thought-provoking points.

"I'm not saying that this has to succeed to be a counterculture. What I am saying is there are plenty of cultural moments that purport a certain kind of anti-authoritarianism or whatever it might be. In every fucking Disney film the baddie is the guy in the suit. I think that for steampunk or anything to actually be a counterculture, it has to be opposing a dominant ideology and I do not believe that there is a dominant ideology that says it's great to live in disposable culture and that therefore there is something contrarian to oppose that. In fact I think the dominant ideology is: 'We throw away too much stuff; disposable culture is a problem,' while disposable culture gets on with existing. So at that level of opposing dominant ideologies I don't think steampunk is doing that."

"That makes sense," I responded. "I think you have a higher bar than I do, but I think it's a reasonable critique."

And it is. When you think about it this way, the question becomes: "What's the real, underlying problem?" The answer, I think China and I agree, is: *capitalism*. That's a bigger fish than we can fry here, but it bears thinking about. Underneath all this stuff, digital and otherwise, is a system that values profit over people and places power above truth. Personally, I think steampunk has a lot to say here, but for now I'll concede the point that when it comes to consumerism, much of steampunk really only addresses the symptoms of an illness that most of us acknowledge we have, but probably don't really think is curable.

"None of this is to say that brilliant work can't be produced in this paradigm," China clarified. "What it is to say is: yes, I do have a high bar. I think if one is going to talk about oppositional culture that the bar *should* be high. I think there are historically specific moments—the hippie culture and the culture of the '60s, of course, you're right, there were all kinds of oppositional things happening—there is no question that the dominant ideology and culture was more obedient, was more right wing, it was more authoritarian, was less open to kind of minorities, sexual minorities, and so on. And the '60s counterculture was a major part of shifting that."

"But it didn't solve the fundamental problem, right?" I said, following his drift. "So in part we are dealing with the legacy of that in today's insipid neoliberalism."

"Absolutely, absolutely." China nodded.

I went on, following the connection through. "We have all these Democrats in power right now, so it's like punching a soft wall. It's a hell of a lot easier to oppose George W. Bush than it is to oppose Barack Obama."

"Absolutely." China repeated. "This is not steampunk's fault. You see part of the problem is as with the argument of mistrust one says these things and a lot of the people who really are steampunk or whatever subculture under scrutiny is, will get really enraged and just sort of say, 'Well what do you think steampunk should do?' I'm saying: 'You do what you want to do.' Make your baubles and enjoy—and I don't mean that pissively; fantastic, those baubles are beautiful. What I'm saying is: 'Let's just be clear about what it is you're doing and let's be clear about what it is you're *not* doing.'

"If you want to make a counterculture, if you want to make an oppositional culture come to the table and say here is why, here is what steampunk, or whatever it might be, is doing that stands in clear opposition to the dominant ideology that is shaping the world at the movement and I will be delighted. Until that happens, it seems to be that steampunk—and this is not a criticism, I'm simply stating it as a sort of an aesthetic and sociological observation—is a subculture based on a certain set of delights in a particular aesthetically interpreted moment, that's fine."

While in some ways China and I agreed to disagree, he certainly gave me a lot to think about. I tend to get into this sort of debate with Marxists, who are absolutely not the right crowd for my "image can be substance" approach. But when you sweep away the ideological dust and look at the clean floor beneath, we concur that real change is possible and that it starts—but doesn't end—with imagining change. And here's where steampunk gets really interesting, because it imagines backward. How can changing the past change our future? Can steampunk actually help us make a better world?

It may sound a bit trite, but sometimes things get repeated because they're deeply true: real change begins at home. With that in mind, it's time for us to tackle...

The Victorian Problem

So God save The Queen!

'Cause everything is possible

For a man in a top hat

With a monkey, with a monacle!

— PROFESSOR ELEMENTAL

"Penny Dreadful" *The Indifference Engine* (2010)

I've spent the past couple years traveling around the world talking steampunk with a huge variety of folks, and I gotta tell ya: it's got a mixed reputation. China is only one among many sharp minds out there that have serious reservations about some of what they see in steampunk. What's more, they've got good reasons. It's no good mincing words here:

The fact is that steampunk's romance with the past can be dangerous. It's altogether too easy to become an unwitting accomplice in the crimes of the past. History has sharp edges.

The attitudes toward technology and change we discussed in the last chapter aren't the only reason these two historical moments resonate with one another. China reminded me that quite a bit of steampunk can be read as glorifying a past time of empire and oppression in a current time of empire and oppression. It's hard to argue with him on this point. The historical parallels are uncanny—so much so that Steven Moffatt didn't have to re-write Doctor Watson to fit the contemporary setting of his BBC TV series *Sherlock*. Just like Sir Arthur Conan-Doyle's original, the 21st century Watson is an injured veteran sent home from an ongoing war in Afghanistan. Don't forget the "dark satanic mills" that Cory Doctorow brought up in Chapter 3, either. Upton Sinclair's "jungle" didn't disappear, it just moved to China (the country, not the guy).

However you feel about today's political or economic situation, there's no denying that some seriously bad stuff happened in the 19th century. You just can't go there without butting up against empire in one form or another. Exploration, adventure, romance and invention are all well and good, but don't go forgetting that nineteenth-century Europe's high society was built on plunder. Imperialism is some seriously dangerous stuff. It isn't enough to say "We want to keep the good and toss the bad." History doesn't work that way. You can't strain the East India Company out of your cup of tea.

Boilerplate sets an example, stepping in to rescue a young woman as the Illinois National Guard fires on striking Pullman workers in 1894 (image courtesy of Paul Guinan)

Y'all Americans (I use the self-inclusive royal "y'all" here) don't get to duck the blame either. What goes for London also goes for Chicago and New York—Cherie Priest has one heck of an imperial foundation for her American steampunk. Ain't nobody off the hook just because the guys in blue killed a larger percentage of the guys in grey. Take a look at the what the U.S. Army got up to in the West after the Civil War (a prime example: Wounded Knee) or at the American conquest of the Philippines, Puerto Rico, Guam, and (temporarily) Cuba in the Spanish-American War. I won't even get started on what the French did in Africa and South-East Asia. Few hands are clean of this mess.

There are lots of things we could call the period of time that steampunk draws upon. So why does the idea of "Victoriana" matter to steampunk? Why do so many of us choose such an explicitly imperial label? What we're talking about, in historical terms, is a time when large scale industry took hold of the Western world. The tendrils of empire had grown into tentacles long before, sweeping up sugar, slaves, tobacco, tea, spices, and markets. By the mid-19th century, those tentacles had gone from octopus to monstrous giant squid, sucking in land, labor, raw materials, and the trappings of power—the epic loot of the ancient world—sacking the tombs of the pharaohs and carting off entire obelisks.

So knowing all that, why slap the grand old monarch's name all over the darn thing? Why sum all that up in glitz and glory, queen and country? Well, sheer romanticism for one thing. Makes sense. It's in what you choose to see. It's just plain fun to imagine yourself presenting your shiny new brass automaton design in front of Victoria and Albert at the Crystal Palace Exhibition. Referring to this age as "Victorian" gives it flair and a romantic character. We are playing, after all, right? Right.

To be fair, Victoria (the Queen, that is) and her ilk were hardly living embodiments of evil. The truth of the matter is that most historical wrongs were committed by people who had at the very least convinced themselves that they were doing *some* sort of good. The 19th century European and American industrial imperial elite—the folks who had the money for exotic clothes and scientific instruments, and the leisure time to tinker, invent, explore, and dream—had essentially convinced themselves that they were embarked on a great project to bring civilization to the world. In fact, many had taken it on as an honorable cause; a grand calling. The White Man's Burden, as imperial laureate Rudyard Kipling, put it, was to raise up the world's dark-skinned savages to Europe's clearly superior standards.

This imperial sentiment was a hallmark of the Victorian Era. It provided fuel for adventure, grist for glorious penny dreadful and dime novel tales, and it gloried in scientific romance. Nothing was impossible, progress was inevitable, and technology would lead the way. Sound familiar? This is the well from which the romance of steampunk draws its water. This glorious imagination can lead to the wondrous world of Scott Westerfeld's *Leviathan*... or to Kipling's crusade. It's steampunk's greatest strength *and* its greatest weakness. Imaginative history can open up a host of possible worlds. It can also attempt to deny the Holocaust.

I love wandering the halls of the British Museum, but I don't for a minute forget about how all that amazing stuff got there. Do I contradict myself? Very well, then, I contradict myself. Who doesn't in this day and age? Take a cue from Walt

Whitman: be large. Contain multitudes. Know the sharpness of each angle of the blade before you pick it up. Those steampunk scissors can cut a gorgeous snowflake out of a folded piece of paper. They can also gouge your eye out or stab a friend in the leg.

I said we'd get to bursting the bubble in the last chapter, and here it is: the point where things turn sticky. We've been flying fast and loose and making a lot of assumptions—in truth, what people usually do. But to speak fully and honestly about steampunk, we have to address the two elephants in the room. Simply put, these are:

1. Romanticizing empire is dangerous.
2. Not all progress (scientific or otherwise) is *good* progress.

Our newfound ability to romp around in the past provides fantastic avenues for the imagination. The Information Age has given us a wealth of material—more history than I can shake a stick at, and I can shake a stick at a lot of history. It's also given us all the tools we need to cut and shape it into our own stories. But we need to do this with some thought and care.

So, how can steampunk avoid getting cut on history's sharp edges? Glorifying yesterday's oppressive empires helps justify today's oppressive empires, and blind faith in scientific progress got us the atom bomb. As responsible human beings in an increasingly connected world, we've got a responsibility to think about how what we say and do affects the people around us. What's more, if we get so high on technology that we forget to question our motives, we might just wind up ending the world.

The rest of this chapter is about some ways in which steampunk is coming to grips with the twin demons of empire and progress. And about how people fit into that equation. Because a steampunk world that glorifies empire or blindly exalts "progress" for its own sake excludes a hell of a lot of people. It's sure wouldn't be an environment I'd bring my daughters into. But I do bring Mimi to steampunk conventions, and will start bringing her younger sister Beatrix this coming weekend at Steamcon 4. Why? Because steampunk isn't just fun—it's smart, self-critical, inclusive, and welcoming. In large part due to folks like Jaymee Goh and Libby Bulloff.

Steam Around the World: Jaymee Goh

> *"There are places between empires where they cannot reach, which are too distant or unimportant to pay attention to. And this is where hidden life thrives, concealed from the powerful eyes but known to those who are curious enough to notice such things."*
>
> **— EKATERINA SEDIA**
>
> *Heart of Iron* (2011)

By mid-summer 2011, I had consumed a lot of steampunk material—hundreds of books, short stories, movies, TV shows, games, and more. I knew that there was a good deal of smart, insightful, and critical stuff out there (along with a lot of not-so-smart, uncritical stuff). And from the conversations I'd already had, I knew that there were plenty of people doing good critical thinking in the community.

I don't recall who it was who first recommended that I check out the internet. I'm pretty sure it was just about everyone I spoke to about the potential problems that arise when you're playing with history—which means a dozen or more people by that time. I was directed to two fantastic blogs: Jaymee's Silver Goggles (*http://silver-goggles.blogspot.com/*) and Diana "Ay-Leen the Peacemaker" Pho's Beyond Victoriana (*http://beyondvictoriana.com/*). "Talk to Jaymee and Ay-Leen" had become a common refrain, and I had no intention of ignoring that advice.

Fortunately, Jaymee dropped by my neck of the woods. She was going to be at GearCon in Portland. What's more, she was hosting a panel titled "Steam Around the World." More serendipity! I got in touch with her right away and set up time for an interview.

It seemed best to begin by asking her what she hoped to get out of a panel on multi-cultural steampunk. The answer was much broader than I'd initially expected. Jaymee is a "shoot for the stars" type—she thinks big and aims high. It became clear very quickly that we weren't just going to be talking about literary settings, but about the steampunk community—and ourselves.

"The potential for real multi-cultural steampunk has always been there," Jaymee said, managing to look serious, concerned, and hopeful all at the same time. "And people have always been dabbling in it. They don't always do it right, but they try."

"What I would like to see happen is not just more people doing multicultural stuff," she continued, "but an actual multicultural steampunk movement going on, where people of different cultures get involved and bring that in. Where I can walk into GearCon and not count myself and my cousin as the only Asians in the space.

I mean, here today I can't see any other visible minorities. There is still this issue in a lot of science fiction conventions where visible racial minorities constitute maybe 2% of the entire attendance. That's a problem, considering that we're over represented in other things like prison systems and drug use."

I could only agree. Where steampunk isn't struggling with this stuff, it should be. Heck, we all should be, steampunk or not.

"I see steampunk as trying to make a movement for inclusivity," she continued, "particularly in forms of fiction and in costuming. But there's an issue with these things too: the fiction, costumes, artwork, and so on are still usually created by a particular kind of person; usually white."

She's right, of course. Though some change is happening as steampunk takes a larger stage and there are some brilliant people who provide shining exceptions to the rule, it remains remarkably ethnically homogenous for a subculture that strives to be as inclusive as possible. So why do we still have these problems?

Jaymee gave me the answer without my even having to ask the question: "Steampunk is a *sub* culture, not a separate culture. Yeah, it's a form of escapism, but we still carry with us ideas and prejudices and stereotypes from a larger culture that we grew up in and we're still trapped in no matter how we try to escape." Remember our definitions back in Chapter 1? There's no escaping culture. Even counterculture carries elements of the broader culture it resists.

The thing about culture is that it's pervasive. Jaymee's right—you can't escape it. Even if you run off and hole yourself up in a cave in the mountains, you bring with you all your assumptions and perspectives. We start learning to swim our cultural waters as infants, and they're our habitat. There's no reset button; no control-alt-delete that will bring us back to base code. The bias is built into the system.

So it stands to reason that when steampunks dig in and play with the past, they bring their own culture with them. And like it or not, western culture doesn't travel light. We carry a lot of baggage, particularly when it comes to the things we've been exploring in this chapter: those big problems of empire and progress. And let's face it, even in 21st century America, we're far from color-blind and a long way from integrated. Sure, we've got a black man in the White House, but as I write this, there isn't a single African American or Native American in the U.S. Senate.

We can't escape these disparities but we can challenge them. In order to do that, we have to bring them out into the light and talk about them. And that's the big insight behind "Steam Around the World"—we *can* steampunk safely; we *can* think about other worlds in ways that don't perpetuate the problems of our own. We just have to actually think about it.

Jaymee explained it this way: "Ay-Leen and I came up with 'Steam Around the World'—a roundtable envisioning a better steam society—because we really wanted to bring up the idea that you can talk about social justice issues in steampunk. And you can talk about deeply philosophical stuff in steampunk that has relevance in day-to-day living. I mean, if you want to bring back ideas from the present into the past, then why don't you think about how that's going to affect the past? And why don't you think about how these issues affect us now?"

Why not indeed? Since they began this endeavor, Ay-Leen and Jaymee's panels have been quite popular. They've presented (together and separately) at steampunk events in Texas, California, New Jersey—and that was just in their first year at it.

Given its intimacy and informality, GearCon was clearly on the open and interactive end of the conference spectrum. So when about 15 or 20 people settled into a hotel conference room to hear what Jaymee had to say, it got real interesting real fast.

When I say "hotel conference room," I mean that in the absolute stereotypical sense of the word. The room had only two real walls—the other two were dividers, behind which lurked other con panels. Lines of slightly plush stackable chairs faced a long table, on which sat a monitor, Jaymee's laptop, and the ubiquitous presenter's carafe of ice water. Jaymee stood in front of the table (a smart choice) and welcomed everyone with a smile. She introduced herself and started by asking a very interesting question:

"*Where* is steampunk?"

This inspired a few blank looks, a couple confused murmurs ("Um, here?" ... "An alternate dimension?"), and what was likely an interested smirk from me. As an aside—proving a point she made earlier—nearly everybody in the room, Jaymee and her cousin excepted, appeared to be white. There was a fairly wide age range, from maybe 12 to maybe 65, and probably a moderate class spread (costumes make it hard to tell, but the fact of costumes itself tends to preclude the extremely poor), but the vast majority in the room sure appeared to be folks of predominantly European ancestry. Getting back to the question at hand...

"I mean geographically," she elaborated, "where in the *world*?"

Now people were starting to get it. "Victorian England" was the very first answer. Another: "Well, more like the British Empire..." More people chimed in with "America, too. The Weird West totally counts." And "Yeah, Cherie Priest and Devon Monk set their stuff in America." "Andrew Mayer's steampunk superhero books are set in New York," another chimed in. "Yeah," said a final respondent, "it's really wherever the Victorians went, right?"

"Okay," Jaymee said calmly, "I'm going to stop you there, because here's the thing: when people talk about what steampunk is, one of the common answers is: 'Victorian science fiction.' And that's one of the problems we're here to discuss. Because even though the term 'Victorian' refers to a time period, an era, it also refers to a particular person who is rooted in a particular geographic region."

"And," she gestured to herself, "that person is definitely not me."

Ah, the Victorian problem applied. This is tangible. Why not use that term? It excludes a *huge* number of people. The majority of the planet's population, in point of fact. Steampunk is large. It *can* contain multitudes, but we've got to acknowledge that there's strong tension between inclusivity and this "Victorian" thing.

"So when *I* hear 'Victorian science fiction,' I kind of cringe," she demonstrated. "Because we are returning to a science fiction that is really very Eurocentric, that is rooted in England specifically, Europe more generally. And it cuts the rest of us out who didn't belong to that geographic region."

She took a kind of conceptual step back and relaxed into a smile. "Here's the thing," she said, "steampunk can be anywhere."

Jaymee's presentation—"Five reasons why multicultural steampunk works"—provides a great backbone for talking about how we can avoid the Empire trap and use creative history intelligently for positive change:

1: *History didn't only happen in Europe.*

2: *Steampunking is about questioning modernity.*

3: *Steampunk is all about subverting histories and creating alternative histories.*

4: *Steampunk is about alternative-ness, not Neo-Victorianism.*

5: *When is steampunk happening? Now.*

Once the floor opened to Q & A, things got *really* interesting. A middle-aged white guy who'd been sitting toward the back, fidgeting uncomfortably for most of the presentation, raised his hand like a shot. Judging from his attire (bought, not made, and pretty standard: goggles, boots, waistcoat) and demeanor (which by this time read as a slightly tense "Why did I choose this panel over the other one?"), this was a weekend warrior—a fan of the fiction there to get books signed by favorite authors and have fun dressing up. I'll stop short of accusing him of "just glueing some gears on it"—he really did seem to care.

Jaymee was polite in noting his enthusiasm, but made a point of calling on a woman to address the first question. She got to him next, but he was already beginning to fume.

"I have an observation to make," he said pointedly. "When you were talking about all the artists and writers who were doing non-western steampunk, you were really excited. You were clearly having fun and it was infectious."

"Thank you," Jaymee replied, not unkindly, but waiting for the other shoe that was clearly about to drop.

"You're welcome. But what was really interesting to me was how your body language changed when you got to your points about racism and colonialism. You stopped having fun, and so we stopped having fun. It was a real downer."

"Racism is a downer," Jaymee replied. "Oppression is not fun."

"So then why focus on it?" he asked. "I mean, we're all here to have fun and play with history and with our imaginations. Why call down a rainstorm on the parade? Can't we just enjoy ourselves without having to worry about all the bad stuff that's ever happened in history?"

"That's not what I'm saying." Her face grew thoughtful as she framed a response she hoped would reach him. "Steampunk is fun. That's a huge part of why we do it. But we can't do it blind. There are other people out there. I'm saying: 'play, but play well with others.'"

"But even now, you're stiffening up," he continued. "That's not fun. That's not entertaining. If we spend all our time worrying about what everybody else might think, we're worrying, not enjoying ourselves. This is a hobby. This is playtime. I don't want the weight of the world on my back when I play—I get enough of that in the real world."

The conversation continued on, and even heated up a bit, engaging the room. The guy wouldn't let his point go. He was determined to hold Jaymee to account for raining on his parade. When the hour came to a close, we agreed to disagree and went our separate ways, but not before Jaymee outlined one crucial point:

"Have fun," she said. "That's great. But don't tune out how it affects other people. If your fun makes other people *not* have fun, then it's not a good game. We have to collaborate. We have to play together."

Play, in this sense, is a lot like freedom. Written rules aside, yours ends where it obstructs another's. This is an important caveat on my steampunk "moving train" (nods again to both Howard Zinn and Cherie Priest here)—If *we* aren't having fun, then *we* are doing it wrong. And that needs to be an inclusive "we." Steampunk is a collaborative project—it's something we do together. We do it together at conventions. We interact with the books we read (remember what China said about the reader's impact on the written word?). And we put our skill sets together to build and operate huge living fantasies like the Neverwas Haul. *There is no solitary steampunk.*

The panel had ended, but hit by a moment's inspiration, I ran after Jaymee to ask one follow up question. "Wait," I said, catching my breath, "you used the word 'collaborate.' I like that. Does building a collaborative past gives us a foundation to build a collaborative future?"

"Yeah," Jaymee replied with an earnest, but now slightly mischievous smile, " I think it does."

Art, Social Change, Martinis... and More Chickens

I'd found the political edge of steampunk—the challenging counterculture I'd seen peeking through all the shiny costumes and airship adventure stories. I wanted to talk to more of the vanguard of 21st century steampunk. So, again through Cherie Priest (I swear, the woman's a living rolodex of awesome) I got introduced to Libby Bulloff, a sharp, edgy, and challenging artist who'd been at the heart of the steampunk scene before there was a steampunk scene. She'd been a part of *SteamPunk Magazine* from its early days and had, along with another local artist, organized one of the first steampunk art shows that I'd been able to track down. Libby's photographs also liberally adorn *The Steampunk Bible*, and her contribution to the chapter on steampunk fashion is a highlight of the book. This lady had been (and, as I found out, still is) on the bleeding edge.

We agreed to meet for a drink at her favorite bar, Vito's, which was a short walk from my place in central Seattle. It was the good season in Seattle—one of the handful of perfect summer days we get each year that make us all forget the grey drizzle that plagues 10 months out of our every year. So, of course, we sat in a dark candle-lit booth. Seattlites... What can I say?

You'd half expect the Rat Pack to walk in the door of Vito's at any moment. It's got a full Italian menu, deadly delicious mixed drinks, a solid selection of single malts, and their bartender mixes a helluva nice martini: Beefeaters, straight up, 2 olives, open a bottle of vermouth in the same room. It's a perfect lair for Libby, a beautiful and flamboyant young woman with blazing red, blue, and purple tinged hair, who counts herself a grumpy old man in spirit.

Libby brought two friends with her: Nathaniel, a kind and soft-spoken guy who I soon discovered was the multi-talented stringed instrument wizard of the spotlight steampunk band Abney Park, and Willow, a one-woman inquisition. We ordered drinks and Willow, a blue-haired, bespectacled hacker as fierce as she's petite, set to grilling me. I didn't expect the Spanish Inquisition, nobody does, but hey—the more the merrier and at this early point in the research I could expense the meal for good material. It was the corporate connection she was worried about—she wanted to know what Intel's interest in the scene was, what my end goals were. Additionally, she was very intent on knowing whether or not they were "taking care of me" as I did this work. I explained how it all started, and how I'd ended up reaching out to Libby.

"I can't speak for Intel as a whole," I continued, "but I know that the people I'm working with on this have their hearts in the right places—they want to learn. I wouldn't have even started on this if I thought they were just going to take it at a surface level and try to market laptops with gears on them or something.

"So, you're not just doing this for free, right?" Willow asks.

"No, my research is paid work," I reassure her. "I'm writing the book on my own time, but I'm not donating any work. I believe in what I'm doing, but I gotta feed my kids too."

"Good," she nodded. "'Cause I'm not about to let you talk to Libby unless this is for the right reasons, you know. It matters how they treat you, too."

This was awesome—smart, kind radicals. My kind of people. We knocked back another round and got down to brass tacks. Willow isn't a part of the steampunk scene, but she is a maker, hacker, and code-wizard who travels around the world working on open-source tech. She's of Libby's crew, and has the gritty fidelity of a true friend about her.

"I really want to get past the image stuff and down into the trenches," I said, emboldened by my second martini. "I get the popular culture angle, but I know there's more to steampunk. I really want to hear about *SteamPunk Magazine*, about the politics, you know, the edge that isn't blunted." I babbled a bit about the Beats and the Pranksters and my own politics.

"For me with steampunk," she said, "I would really like it to be a community, a family of people who all believe in the same things. At the moment, it has some political nature behind it, but a lot of the individuals who are currently active in steampunk I think are mostly involved in a scene. I'm trying to push it toward being a community."

"And how do people respond to your pushing?" I asked.

"They hate me." She laughed.

This was gonna be a three-martini interview. Fortunately the walk home was downhill.

"It's really kind of difficult sometimes to get people out of that comfort zone and say, let's talk about the social ramifications of what you're doing, or the political ramifications, or what you're saying about gender when you do what you do. Or how are we affecting our environment around us?" She looks at me knowingly. "And it makes people itchy; they sometimes don't want to listen. They're just like, 'You're strange. Go away. I just want to buy a cool hat and have goggles on it.'"

"And that's okay, too," she added. We tucked into fresh drinks. "But for me and for my crew of people in steampunk, we've been really trying to get people to think about the bigger picture, and think about how steampunk affects other things besides just the people involved within. Our larger communities, our political structures, our environmental structures, and technology specifically."

"So it's safe to say you see some potential in it for change," I asked. "Effective change?"

Libby nodded. "Yeah, definitely effective change."

"How does that work?" I pressed on. "Where does the change come in?"

"Well, if this does continue past being sort of an aesthetic fad—which is really big right now—I'm hoping that we can use this to create bigger and better technology, to help our entire community outside of the scene, too." She pauses, and then continues, saying, "Steampunk has the potential to really grow open source and maker culture if people let it."

"So how do you do that?" I wondered.

"Well I know a number of people in steampunk and in other subcultural scenes who have started makerspaces, or attend hackerspaces and things like that. When you get to interact with people of all ages in these spaces, you get to teach them, and then they teach you. And that kind of exchange of knowledge is how we all better ourselves. It's really great to walk into a makerspace full of folks and say,

'Hey, this is what I'm working on. What are you working on?' Teaching doesn't have to happen in a school. It can happen anywhere. I'm really impressed at a number of the people in the steampunk scene sharing their knowledge, and not being proprietary about their projects."

And maker culture comes up again. Something that definitely calls for further inquiry. We'll explore this essential piece of the puzzle soon. For now, though, I have an oddly compelling urge to loop back to my original question again, asking, "So what is steampunk? I mean, how do you define it."

Libby leans over and shoots me her trademark look. "You tell me, Mr. Cultural Historian."

Touché. "Well," I said, "my daughter will tell you that it's about chickens and teapots."

After the laughter subsided, we agreed that was as good a working definition as any. We all, as Libby tweeted, "grokked the clock-work bawk-bawk."

I kept my mouth shut so she'd keep going with this thread. "It's been very hard to explain to people that yeah, steampunk is political. Any sort of subculture is. We're causing a riot. When you walk down the street in a top hat and spats, you are causing a riot. You're making a statement, and it is a statement about politics, and gender, and society. "

So the historian flares up again. I gotta know—she mentioned top hats, spats, and riots in the same breath. "Why all the Victorian stuff, then?" I blurted. "I mean, top hats and spats sure as heck weren't rebellious a hundred and fifty years ago, right?"

"Well, first, I like the aesthetic," Libby admitted. "I liked all of the various types of tailoring in the garments, and how heavily-tailored a lot of the garments were. Additionally, the same amount of detail that was paid to the fashion of the time—the heavy embroidery, the bustles, the hems, the varying lengths of gentlemen's pants—was paid to the technology and to the architecture of the time.

"But you're talking about a very tough time for most people." I had to play devil's advocate on this one. "The people who had access to that in the 19th century were very wealthy, white, and usually male. And yet, you're trying to tap into that to do something very different with it. Why reach to something with that kind of a social heritage?"

"Well," she shoots me that trademark stare again. "You say: 'Why reach to something with that kind of heritage?' And I say: 'Why not?' I mean, yeah, it's really tragic to look back and think about how many people were living in poverty, and how many women were oppressed and had no property or right to vote, and what it was like to be a person of color, or to be gay. And that was horrifying. But now we live in an era where it's maybe not super simple to be of a non-white, heteronormative, standard gender identity, but things are getting better."

Libby looked at her friends, and said, "So I think that is very interesting for a lot of people now, because we do have the freedom to be of color, and nonreligious, and gay, and gender queer, and things like that. But at the same time, we're not necessarily given the freedom to level with our tech and level with art. And so I think for me, it's really kind of interesting to sort of take what I can from the past and leave the rest. And then add it to the future, and then hope and pray and cross my fingers and see what I can do in the meantime to make it better."

I was genuinely inspired. "How do we keep that sense of possibility without romanticizing the Victorians?"

"I don't know. Just do it," she snorted. "That's what I always tell people when they say, 'Well, I'm going to need this other thing before I can do this project. Or, I just don't think it'll work...' I don't care. Get up and do something."

She was on a full roll by this point. "Imagination and politics have to go together, because otherwise what's the point? We have to create change. And in order to create change, you have to envision what you would consider to be closer to perfection. And that involves using your imagination, because we certainly don't live in a perfect place. I don't want a utopia by any means. I just want a place where people can live, and be a little bit more creative in doing so."

Imagination and politics have to go together. Talk about lessons learned from steampunk.

Libby continued. "One of the biggest things that I have sort of pushed as far as my agenda as a member of this subculture—this movement—is the idea of 'I have met the enemy and he is us.' We have to casualize steampunk if it's going to live past being just sort of a fad that lasted a couple of years, and was eaten up by the media."

"So it isn't enough to just free your mind?" I asked, slightly tongue-in-cheek.

"No." Again, the look. This was serious business for Libby. "You can't just free your mind and put your brain in a jar. That's useless. Then you have a brain in a jar. Or you have an uploaded brain in an iPod, and nobody cares, because there's no one there to listen."

We all went a bit silent at that point. Eventually, the conversation wound down, and Nathaniel mentioned an all-ages Abney Park show coming up—it seemed the perfect opportunity to include my kids. I couldn't wait to introduce Libby to Mimi —a sharp, fashion-minded rebel of a kid if there ever was one.

An Evening With Abney Park

Nathaniel Johnstone was pretty quiet over our drinks at Vito's, but that'd've been a very different story if he'd had a mandolin in his hands. Nathaniel is a crazy-talented musician who plays with the showcase steampunk band, Abney Park. I'm not entirely sure there's an instrument he can't play. The things he can do with a mandolin, banjo, fiddle, or guitar are just mind-boggling. As we left the bar, he kindly invited me (and Mimi!) to be his guests for Abney Park's next show. It was a huge honor to say the least.

Time Travel Note

Nathaniel has since left Abney Park to kick up his own project, The Nathaniel Johnstone Band. Libby plays "the iPhone and other assorted percussiony bits." I've seen them in concert, and they're so culture-punking awesome I went straight back home and blogged the heck out of that show (*http://culthistorian.com*). To me at least, there's an irresistibly profound cultural beauty in a squeaky-clean segue from gypsy surf rock into the early '80s classic "Hey Mickey!" They never even missed a beat for the belly-dancers.

The concert was an all-ages show just south of my neighborhood in Seattle's SoDo (South of Downtown) region. It's a warren of warehouses and contractor supply stores, peppered with oddments like the Safeco Field (home of the Seattle Mariners), Starbucks' corporate headquarters, some strip clubs, and a few musical venues. Libby agreed to meet us at one of the latter, where Abney Park was playing that night.

Mimi, then 8, was endlessly curious, fascinated with performance and costume, and took after her father in being inescapably drawn to the fringe. She'd become, in her younger sister Beatrix's words "b-sessed" with steampunk, and knowing that there was space for both young and old in the scene, it was high time I brought her along. She was thrilled beyond speech, and nervous as heck as we queued up to claim our complimentary tickets and find a good space near the stage to catch the show.

Nathaniel Johnstone and Captain Robert of Abney Park perform at the Seattle Steampunk Soiree, July 9, 2011. (photo courtesy of Byrd McDonald)

Libby found us in line. True to my expectations, her multi-colored hair and brilliantly eclectic style captivated Mimi, immediately validating all of her crazy fashion fantasies and, I'm fair certain, filling her inspiration tank for some time to come. She beckoned us to follow her.

I'm not sure to this day whether I warped Mimi for life by bringing her along, because she now thinks that going to a concert means waiting in line for a few minutes until a friend who knows the band comes out to bring you backstage to hang out in a little room full of couches, makeup, and free soda where you chat with the band for an hour while the opening act warms up the crowd.

When the lights went up and the band hit the stage in a blast of steampunk fervor, Mimi was transfixed. Heck, I was transfixed. I have to confess that I didn't think all that much of the band's music until then, but Abney Park puts on one hell of a show. You know how you can look at a person from a distance and not find them all that attractive, but then when you sit down and talk to them—hear what they have to say—they're all of a sudden gorgeous? Well, this was kinda like that. It took the visuals—the pure spectacle of it all, to clinch the beauty of what they do. I could go on about the inter-dependence of image and substance *ad nauseum*, but I won't. Not because I don't want to (far from it) but because it would detract from what I really want to say here: Abney Park just plain blew my socks off that night.

While Mimi became more and more absorbed in the show ("Daddy, I can *feel* the music!"), I was taking in the whole scene. When Abney Park puts on an all ages show, they really mean it. The bulk of the crowd was probably anywhere from their late teens to their mid-forties, but there were more than a couple young kids, and no shortage of older folks as well. I saw a variety of genders—male, female, indistinguishable, and a few people of color (this last, as we know, remains a challenging space for steampunk culture). Costumes and props abounded, but were clearly not a pre-requisite. Vendors hawked wares—handmade, second-hand, and mass-produced. Aerialists performed between sets—underlining the 19th century circus/carnival atmosphere that many steampunk performances have come to spotlight.

By halfway through the show, Mimi had made a friend about her size, and they were bopping up a storm along a crowd barrier to the side of the stage. She made it through nearly two entire sets, and I have to say that I was impressed. I'm not sure I would've made it to midnight at a rock show at the age of 8, but she did. I didn't even hear an "I'm tired" until about a quarter to twelve.

On the way back home, she declared her intention become an "artist, rock band fiddle-player, airship pirate." I heartily endorsed this as a career direction, and we yawned through a fun chat about all the changes she's going to make in the world once she's famous like Libby and Nathaniel. Girl's got scope and is gaining perspective by the day. She hit her second (okay, maybe third) wind when we arrived back at the apartment, but passed out soon thereafter with her head stuck in a reproduction 1897 Sears Roebuck catalog. My daughter—asleep on the past, dreaming of the future.

Mimi, transfixed by her new "favorite band ever!" (photo courtesy of Byrd McDonald)

Punking Time in Key West

It turns out that steampunk is just one of many shiny spotlights trained on a changing world. James, Byrd, and the documentary crew travel to Key West Florida to look beyond steampunk and discover a wider variety of people who are all "punking" time to design a better future.

Speaking of time, we are in the midst of a very interesting one. Artists and writers all over the world are imagining and playing with the past and the future in the attempt to create a new present. This is not new. What *is* new is that fact that they are doing it *consciously*. They realize that playing with the past and imagined futures can actually have an effect on our real future. These people are all makers, hackers, and culturepunks of a specific kind—they are striving to build a different world than the one we live in today.

The Key West Literary Seminar is hosted by James Gleick and Margaret Atwood. You couldn't gather a more heady collection of literary giants. James and the crew spend a dizzy few days talking with Margaret Atwood, William Gibson, and China Miéville. (In the spirit of punking time with China in Chapter 8.) The crew learns that the desire to hack and change the past, present, and future is not just for steampunks.

Living in the Future

James H. Carrott (Key West, Florida)

> *I apologize in advance that I'm reading from my phone, but hopefully you won't notice a difference, because we're living in the future.*
>
> **— DIANA KHOI NGYUEN**
>
> Key West Literary Seminar (2012)

In January 2012, a group of writers, thinkers, and literary enthusiasts gathered together at the southernmost point in the continental US for the Key West Literary Seminar. Not a bad place to be in the deepest throes of winter. This year's focus was "Yet Another World: The Literature of the Future," an appropriate topic for our inquiry if ever there was one.

We heard about the event from China Miéville. He was having tea with Brian in London when they got to talking about *Vintage Tomorrows*. China was good enough to make the introduction and the good people at the conference were angels and allowed us to come. Brian couldn't come because he had to go to the Consumer Electronics Show in Las Vegas, so I headed south with Byrd and the documentary crew in tow. (Brian was heartbroken.) The speaker lineup for the conference was incredible. Some of the best minds in the world had come to Key West, and we had open access and no end of questions.

Key West wasn't what I expected. I walked off the plane onto the sunny tarmac of an airport from an earlier age. It was decorated with an odd blend of Christmas decorations and a sign that proclaimed: "The Conch Republic only 90 miles from Cuba!"

The island was oddly appropriate in itself. Key West was full of unexpected 19th-century glamour, but it was also whored up for the tourists. It's mojito-ville, but if you look deeper there's something real there, something with roots that connect it to a tangible past. I sat in a bar called The Green Parrot and collected my thoughts before my first chat. A sign on the wall proudly proclaimed that the place was over a century old, founded in 1890. Souvenir t-shirts displayed for sale proclaimed "Excess in Moderation!" and "Parrotphenalia." And yes, I actually did order a cheeseburger in paradise (sorry).

From the event's opening moment it was clear I recognized that the "literature of the future" was actually very much the literature of the present. But this present was uniquely conscious of cultural change. The writers gathered here were not futurists. In fact, writers of science fiction *per se* were distinctly in the minority. But, as Program Chair James Gleick put it, "What they do share—what their work reveals—is a deepening awareness of past and future, which also means an awareness that our world is not the only one possible."

OUR WORLD IS NOT THE ONLY ONE POSSIBLE: MARGARET ATWOOD

I hate being late to things. Being early helps me feel more prepared. Showing up early to this event (mid-day Thursday) turned out to be the right call. I bumped into a PR person from the Key West Tourism Board and chatted a bit about the event and why I was there. She asked if there was anything she could do to help.

I acted quickly with a firm, "Yes!" I knew that Margaret Atwood was going to be insanely busy all weekend playing ringmaster alongside James Gleick. "There is something. Do you know if Margaret Atwood has arrived yet?"

She checked her phone. "Yes. She'll be here for a rehearsal this evening."

"Do you think she might have a half-hour or so between things to speak with me?" I asked before my nerves could get the better of me. "I don't need anything formal, just a little time with her and my voice recorder."

A brief phone call later and it was set. Margaret and her friend (and fellow brilliant writer) Valerie Martin would be there in about an hour, and they'd be happy to speak with me in between their obligations. All of a sudden I found myself sitting around a table on a beautiful Key West evening with two of the most brilliant women I've ever had the honor of meeting. There are few people on the planet who know more about "possible worlds" than Margaret Atwood.

I started off by explaining a little about *Vintage Tomorrows*. "So," I said, getting to the point, "it's starting to feel like the 'steampunk is the tip of the iceberg' analogy isn't enough. I think maybe it's more like one island rising out of a volcanic archipelago, and I suspect there's something much bigger bubbling underneath it all. What does it mean to look backward so we can look forward?"

Her answer went straight to the root question of... time travel! "It's a very difficult thing to make totally plausible as you know," she started. "There have been various people quite early on who fooled around with time travel. H.G. Wells was the first that anybody knows anything about as far as I can tell. However, he was travelling into the future. John Wyndham has the story in which people are annoyed to find that travelers from the future are drifting by and sightseeing. They are invisible but they're coming into people's living rooms and observing them. But that's all just visitors from the future," she wiped the words away with her hand.

Program cover, 30th Annual Key West Literary Seminar: Yet Another World, 2012. The image is from a set of cigarette cards by French artist Jean Marc Côté, who was commissioned in 1899 to draw pictures of life in the year 2000.

Margaret Atwood during her panel at the Key West Literary Seminar (photo courtesy of Byrd McDonald)

"But the idea of going *back*," Margaret continued. "This idea of going back is different. Are you allowed to go back and alter anything? If you alter one thing, is that going to alter everything else? What if you change things by the mere fact of inserting yourself into the past?"

"Well," I said, "for steampunk I think it's less about trying to create a plausible narrative or explaining why this would be possible, and more about living in an alternative past, in a world that could have been different."

"That might have been different, yes." Margaret paused for a moment in thought. The sheer quantity of relevant and useful information in this woman's mind is staggering. She claims that this is just because she's "really old," but I'm not fooled for a second; she wields her knowledge like a virtuouso, conducting an orchestra of ideas around every point. "So there are a number of early stories or earlier stories like that that deal with those kinds of time travel and a lot of the early utopias include looking backwards—they involve getting into the future by the central character having some kind of comatose state. This is what Woody Allen was parodying in *Sleeper* where he climbs up covered in tinfoil and finds out that he's in the future.

"So looking backwards," she continued, "it's a Sleeping Beauty thing. And sleeping beauty is in fact a time travel story, so is Rip Van Winkle. So are the old stories about the fairies—that fairies came along and offered you a party and a good time and you went into their dwelling and thought that you were spending three nights and then you came out and everybody you knew was dead and it was 100 years in the future."

"Like how time works differently in Narnia?" Man was I out of my league here... and I thought I was a serious SciFi/Fantasy geek.

"Yes, time is different in the other world," she agreed. "And when you come back, indefinite amounts of our time have passed."

We switched gears then, and talked a bit about changing perceptions of science. She reminded me that skepticism about "progress" was hardly a new thing. This, of course, brought us around to the changing image of the "mad scientist." They're certainly frequent characters in steampunk, and what's more, mad scientists provide a perfect example of looking to the fringes of culture to take its pulse. How we feel about science absolutely shows up in how we depict genius, particularly of the lunatic and improbable sort.

"I place the origin of the mad scientist with Jonathan Smith, in book 3 of *Gulliver's Travels*," she said. Margaret is one of those people whose every word comes out definitive. The odds are high that she's already researched and written extensively on just about anything you might be talking about.

"You had alchemists before that time," she went on, "and Dr. Faustus figures. But number one: there weren't any scientists yet, and number two: they weren't mad—they had sold their soul to the devil (which is a different thing) in exchange for worldly riches and power. They weren't crazy, and they didn't want world domination. Faustus probably could have gone for world domination but he doesn't, he goes for wine, women, and song and having a good time. He's actually quite a nice guy. When you think of the things he actually does, you think: 'Well where's the harm in that? He wants to live more.'"

And there the conversation side stepped a little. She started with, "I don't know how many comic books you've read..." Here I had to laugh, as I've been buried in comic books both personally and professionally for decades. "...but in Captain Marvel in the 1940s there was little character who was actually an alien from outer space in the shape of a worm. His name was Mind and he communicated through a voice box because he was such as little worm." I knew Mind well. My swan song as the brand manager for HeroClix (a superhero miniatures game) was a Golden Age DC Comics set that included a figure of that endearingly maniacal little space invertebrate.

Margaret continued, "Mind wasn't a mad scientist, but he was a world domination guy because of his prodigious brain power. I did a speech for a neurologist's convention a little while back about the brain as a character in literature. People's interest in the brain is actually quite recent."

The brain itself as a character. Now this was an interesting place to start my literary adventure in Key West.

Granted, these are the kind of conversational paths one ought to expect when speaking with Margaret Atwood. There's a reason this woman is a legend in her own time—one which became clearer and clearer as a conversation that started with time travel developed into a discussion of the human mind itself. This was far from a tangent; she was tying it back around into a brilliant bow for me.

"The Greeks weren't interested in brains much at all as an organ," she continued, "I believe they thought that thinking happened in your liver or something. The Chinese weren't really interested in brains, either. It's only relatively recently that the brain comes in. You can find it in Shakespeare to a certain extent, though he's more interested in the heart. It really comes in in the 19th century, when people got interested in nerves and neural wiring and they got interested in galvanizing."

Wow. It turns out that our fascination with the human brain itself is also a creature of the age from which steampunk digs up its inspiration.

"That's where you got Frankenstein coming to life through electricity and this wiring that was connected to the brain. People in the 19th century got very interested in that, so the brain and the nervous system became something that people wrote about. That's when you started getting detached brain stories: the brain all by itself controlling things , brains that are operating on their own."

Ah, the brain in a jar. An apt subject for contemplation in an age of robotics, microprocessors, and intelligence-mimicking algorithms. Just what is it that makes us human? What is our relationship to reality? Are there limits to what we can create? Just what kind of a character is the human brain, anyway? My own mind was spinning.

I tried to regain what poise I could muster. "What you're saying about our interest in the brain," I said, "brings me to a bigger question I've been pondering, which is whether there's something happening in our culture right now, the effects of the Information Revolution perhaps, that's similar in some ways to the change that was happening around the Industrial Revolution."

"More like the Johannes Gutenberg event," Margaret responded, "which was early Industrial Revolution." Right. I am again reminded that I'm talking to someone with a sense of time even broader and deeper than my own. Of course, the technological changes we are experiencing now are on a scale with the print revolution Gutenberg kicked off in 1439 when he put together movable type with a printing press.

I had the chance to squeak in o... ...re question before we had to head out to the opening night talk and ... ...ermined to take this all even further, I brought up the ide... ...h intelligence, which I'd been thinking of as an... ...ry age. "On this topic of technological ar... ...eep pace with one of the most nim- ...ugh the work of a lot of the authors ... of the consciousness of cities or

...pping my question with bril- ...r, she cut a shoot ripe for ... consciousness of stones." ...One Ring, though that's

... have numinous pow- ers ... ld numinous stones, of w... ...ke really, really old. Some... ... plant and a stone —and ... ...se souls of other beings w...

"So o... ...fic part of the 19th centu... ...s, things be- came thing... ...oved from animals and i...

Wow. I ou... ...but they somehow always ... ...mind- ed of something f... ...d me when I was a kid ca... ...ere by a witch, was ticki... ...hat was ticking away the hours

Margaret picked rig... ...n interesting thing has happened to time and ho... ... the invention of the clock, the sundial was the thing that measured time, and time was seasonal—it went according to the movement of the sun and the shadow pass, etc. Then the clock came in and time became divided into measurable segments, so the clock became the metaphor for time; it becomes a mechanical image.

"Now in the age of the internet, we're losing that clock metaphor for scrolling—it's scrolling that moves through the internet in that form that you get on blogs. It's something that is pasted up at the top and the scroll moves down, or if you watch a hot item on Twitter the scroll is just going down all the time. It's moving much more to being a stream, so time *streams* by now instead of just ticking by."

These mental images—clocks and streams, numinous objects and virtual metaphors—stuck with me as I sat, later that evening, in the building's theater, listening to William Gibson respond to a question from Douglas Coupland: "Is artificial experience lived experience? Does artificial memory count?"

"I am inclined to suspect," Bill replied, "that the thing our great-grandchildren will find quaintest and most inexplicable about us is that we insisted on making that distinction. I literally don't think that they will be able to make the distinction between what we still quaintly call 'the virtual'—as though it wasn't real—and the 'real.' I think that's well in process of completely dissolving and becoming as mysterious as my imaginings of what it was like before audio recording was possible."

"The historical evidence," he finished, "is that imagined literary futures melt like ice cream in the trunk of your car. As soon as it comes from beneath your quill pen, it begins to acquire a patina of quaintness, that eventually becomes totally fantastic."

Yet again, my mind was well and fairly blown. We're clearly sitting at the cusp of change here in Key West—and that change has something to do with the nexus of information and time. As we scroll into the future, what will we look back upon? How will imagined *pasts* hold up in the trunk our future car?

While I find it hard to imagine a hairless future smirking back at its quaint hairy ancestors ("Why would they have cared about physical stuff? How droll. Pass the e-crumpets, would you, old chap?"), I do see bits and bytes taking shape around me nearly everywhere I turn. These luminaries were on to something. I guess that's why we call them luminaries.

THE INFORMATION: JAMES GLEICK

As Margaret Atwood so aptly pointed out, and Bill Gibson underlined in deep felt-tip metaphor, our world is changing, more rapidly each day. I caught renowned author and conference co-chair James Gleick for a few moments after a particularly engaging panel and asked him what inspired the collection of writers and topics at the conference. His response underlined a sensibility that has resonated throughout our research:

"Some of the things I feel connected to as a nonfiction writer, my personal interests," Jim said, "are technology, information flow, the rapidity with which the world is changing—these have all been themes for me in my nonfiction writing and they are both the themes of this kind of fiction and the reason that this kind of fiction is becoming so important, so interesting. The reason that somebody like William Gibson is not a marginalized science fiction writer, but a kind of prophet for our times, is that his concerns are the issues that matter to those who are most urgently caught up in the way the world is changing."

The future is nearer to us now, and is approaching—as Jim Gleick has noted in his writing for over a decade—increasingly Faster. While his work initially examined a wide variety of expediting experiences, from fast cars to fast food to things like "close door" buttons on elevators, it soon became clear to him that "what mattered were technologies of information... fax machines and email and the Internet." In the bigger picture, he observed, "all of that stuff is what really affects us as humans."

We talked about how steampunk plays into this picture—as a place where information technology helps facilitate our ability to question and process the very changes it is bringing about. Our cultural sense of time is changing. More importantly, so is our sense of our place within it.

"It's about a kind of time travel, right?" he said. "It's about juxtaposing different times and mixing the future with the past, which I think inevitably comes from a couple of things: one is the sensation that some of these panelists were just describing of whatever you're trying to write about, you're afraid it's going to be obsolete in three months. You try to describe some imaginary future and the future arrives before you're even done.

This sense of "time travel"—of living in the future or constantly having the future catch up with us at the turn of a corner—is central to the cultural impulse that fuels steampunk. This is bigger than gears, goggles, or even a deeper techno-anachronism. It's about our relationship to information, to other people, and to the world around us. It's also about our relationship with ourselves.

The steampunk panel at the 2012 Key West Literary Seminar. Pictured (l-r) are Robert Krulwich, China Miéville, William Gibson, and Dexter Palmer. (photo courtesy of Byrd McDonald)

Steampunk Parallax: William Gibson

When you're covering an event like this, you keep your eyes open for unexpected opportunities. Byrd and the rest of the documentary crew headed off for some prescheduled interviews while I hung back. It's a skill I learned as a radio reporter and have employed to great effect ever since. It's not so much stalking as an attentive lurking, and it paid off in spades just before the conference's steampunk panel. Yes, a steampunk panel at a conference on the Literature of the Future. It sounded kinda crazy at first, but it's already making a ton of sense—ya gotta look back in order to move forward. Or at least to move forward with a purpose.

I'd been hanging around the lobby and noticed William Gibson heading upstairs to the green room with Robert Krulwich, an NPR host and commentator who'd been chosen to serve as the moderator for the panel. Gathering up my guts —in the immortal words of *Zombieland*'s Tallahassee (Woody Harrelson) it was "time to nut up or shut up"—I approached Bill and asked if he/they wouldn't mind me joining them as a "fly on the wall" while they prepared for the panel discussion.

It was a fascinating conversation to say the least. Krulwich is a popular science guy, likely chosen because of the subject's recent surge into popular culture.

He didn't know beans from steampunk—which was fine on a couple levels: first, despite how it may sometimes seem to those of us tuned into the scene, steampunk is not in fact omnipresent in mainstream culture; and second, you don't need to be an expert in something to ask good questions or to drive a good conversation.

Credit where it's due, Robert had done some homework, reaching out to Bill's former collaborator Bruce Sterling, co-author of *The Difference Engine.* (Brian catches up with Bruce in Chapter 17). What Bruce fed him, however, was, at least to Bill and me, pretty far off the mark. At least in emphasis: ominous things about "preparing for the death of our tech" and "a way to be angry at big corporations."

Bill sat back with a laugh and said, "steampunk strikes me as the least angry quasi-bohemian manifestation I've ever seen. For god's sake, it's about sexy girls in top hats riding penny-farthing bicycles. And they're all sweet as pie. There's no scary steampunk."

I grinned. Bill had the right of it—in emphasis and counter-balance at least. Sterling's not flat-out wrong, but he's off on flavor, on the overall, underlying *feel* of the thing. Of course, he knew it was more complicated than the sexy girls on penny-farthings (though we both enjoyed the image).

He went on: "Actually, I think my favorite steampunk things *are* scary steampunk things though—some kind of strange guy with what I guess is an assumed Anglophone name in Bulgaria who has a studio and they make steampunk bondage masks and sell them on eBay. Cory Doctorow buys himself one every time he completes a book. And they are incredibly dark, spooky things: full head masks that you have to like lace something back and they're made of horse hide and brass and repurposed industrial bits and they are really formally fucked up and interesting things."

Here's where I have to envy the heck out of Brian getting to hang out in Cory's office and play with his toys. Well, I get to be here and he doesn't. So there.

Robert asked a logical follow-up, "So where does this leave you? If it's not angry and it's not youth identity experimentation, and it's not whatever Sterling said... Then what is it? Why is it so popular? Or maybe you don't even think it is?"

And Bill demonstrated his wisdom: "Well, I'd rather leave it as a question than have half-assed answers."

Which is not to say that all Sterling's answers were half-assed, but rather that Bill Gibson understood something fundamental about steampunk: it's a cultural work in progress—part of a set of open-ended questions we're asking ourselves about our technology, our selves, our futures. I've heard hundreds of answers to this question—many of them my own—and they're all shades of grey. (Though mine are a very clear and light shade of grey, casting few shadows and adding mostly enlightenment. But I digress...)

We talked a little about how steampunk has been kind of splattered across culture. That there's some brilliant, edgy art, and a lot of DIY stuff that runs the range from inspired to schlock. There's pop culture creamsickles, spray painted Nerf guns, and krazy-glued junk piles. Heck, I've got half a shelf of books on how to make steampunk plushies, how to draw steampunk characters, and the list goes on from there.

But here's where it got truly interesting; the reason I'd come to Key West in the first place, and why I was headed to Australia the following month. I asked Bill what he thought about all that "stuff," and he kinda brushed my question aside in favor of a deeper point.

"Well, there's something going on," he agreed. "There's something wider going on culturally that I don't identify with steampunk, but I think steampunk might be another sort of slightly more exotic symptom of it."

I couldn't suppress a grin. Bill-freaking-Gibson (I know that's not his middle name) had just, without prompting or a direct question, affirmed the suspicions I'd voiced from the start of this project. There's something bigger going on. And that's what this chapter is really about—the something bigger.

Bill went on to call out something really interesting. Another not-steampunk, but related, "symptom": "Young men all across America, if they can afford it, and if they understand the status codes of their age, are wearing nice, meticulously crafted, very expensive reproductions of 1900s work wear. And that's like an enormous industry and that sort of thing interests me, for whatever reason, a lot. And I've gone and met the people who do it. I know some of the people that design it and the people that make it and what the feel is."

"You mean like working class clothing?" I asked. Some steampunks go in that direction, but very few—steampunk fashion is really much more about dressing to the nines in flamboyant style.

"Yeah. Yeah. There's a guy in San Francisco named Roy. Roy makes Roy's Jeans in Roy's Apartment and he sews every stitch. He does everything. It's a one-man operation and they look totally like classic straight leg 5-pocket button fly jeans on the outside. When you turn them inside out, they'll astonish you, like the other stitching is in like rainbow colors and the pocket bags are bright red and it's got his name hand-embroidered on them. It's this crazy thing and it's very, very zeitgeisty. It's sweating *zeigeist.*"

A good interviewer, Robert pushed: "So, why?"

Sometimes not the most patient interviewer myself, I jumped in: "There was an arts and crafts reaction to the Industrial Revolution, too. There's a kind of sense among people who do steampunking about craftsmanship and artisanship and how all that was lost. Of course, there's irony there because when steampunks talk about it, they are talking about the time of the industrial revolution."

Bill saw an even deeper pattern: "Well, it isn't actually the first time it's come back. You should have a look at the British Neo-Romanticists. It was immediately post-war. It was very strongly that. But what happened was, during and after the war, this thing was coming up in England and it was really British—that was part of it. It was a really British kind of retro modernism and then abstract expressionism came through and... like a glacier, it plowed the landscape flat. And it just killed British neo-romanticism."

Man I love this guy. Further confirmation that this is a post-war thing we're talking about. (Also, more reason to hate the abstract expressionists. As if I needed more.)

We talked some more about the kinds of craftsmanship that were disappearing in the post-war British world. Bill mentioned "the guys who had little shops who carved the ponies and mythological animals for roundabouts. And traditional tattoo artists. And people who decorated cakes..." Then his eyes lit up again.

"The other person you can't—you know about Humphrey Jennings? You can't write about what you're writing about without reading *Pandaemonium.*"

Bill went on while I scribbled wildly in my notebook. "It's this astonishing book that is a collection of little short bits that Jennings lifted from first person accounts of encountering alien technology. And my favorite one—Bruce [Sterling], of course, turned me on to this, he actually just sent me the book—is one where this curate goes to a garden party and hears an Edison phonograph and he comes home and writes in his diary and he's like sweating fucking bullets. He's enraged, he's terrified, he's heard a dead woman speak, he feels like puking. I thought, 'Fuck, this is amazing!' And it's all like that, stuff like, 'I rode the steam train yesterday. It attained a speed of 8 miles an hour. I think my body may never recover, all my joints are aching.'"

"What I can hear in that, too," I said, "is the voice of people complaining about their cell phones."

"Yeah, it's exactly that," he agreed.

"Everybody can see me from space!" I mimicked. "Shit!"

Robert rejoined the fray: "So, underneath all this, is that what you'd like to recapture? You'd like to recapture what it's like to be astonished by a culture's inventor in the moment?"

"No." Bill leaned back a little. "I think its value mainly consists of providing parallax. You know, if you lose an eye you have no sense of depth. But when—as a reader or writer—you make a structure of a rigorously executed recursive novel, say, about technology, then you have two eyes and you have the provision of depth when you consider history and when you consider your own situation."

Robert, the mainstream reporter: "Is that a fancy way of saying you have the eyes that now know what it's like to get on the train, but you keep the eyes that didn't know? That you hold on to a feeling for surprise? Is that the depth vision?"

Bill (kindly) declined to be reduced. "I don't know." He said, "I mean, there are some things that can only be said in a very fancy way, unfortunately. I mean, common sense is largely worthless for the understanding of certain concepts. You can't be too reductive with this stuff or you lose the potential for meaning.

"I mean, I'm not really interested in discussing things that I fully understand," he continued. "I don't know what those things would be.I'm not a didactic writer. I'm writing to induce the reader to ask questions—to leave the reader with doubts, essentially. Often when the readers come back to me and ask questions, sometimes—well, not often—sometimes I actually discover in the course of doing that I found sort of an answer."

Robert followed up brilliantly: "What's the best question you can ask about this subject?"

Bill thought for a second. "Well, 'what's it good for?' Which I think I just addressed. I mean, when we're talking about brown clothing with brass accents, probably not very much in the long run. When we're talking about literary strategies that allow us cultural parallax, I think we're talking about something that's potentially very valuable."

Here I had something to add: "I've been playing around with this for a while," I said, "and the metaphor that I use sometimes is from *Pattern Recognition*—when Case describes jet lag and the stress of flying across the world as feeling like she's left her soul behind; it hasn't quite caught up to her yet. I feel sometimes like that's what's going on culturally around us—it's this sense that technology has progressed so quickly and that all this shit is happening so fast that our soul hasn't quite caught up to us yet, that we can't quite trace it back to who we are. So we have to invent these stories and we have to build in this artificial history to it and create myths so that it has a marker and makes sense in our story. Otherwise we just feel a little bit like we're spinning."

Bill made my day again. "Yeah," he said, "I agree with that. I think part of the problem with starting with steampunk as a signifier is that it's actually only one signifier in a field of signifiers.

"Of all the more impressive examples of what you've just described, the one that most comes to mind for me now couldn't possibly be identified as steampunk because they're not focused—they're not Victorian in concept, but they're doing exactly the same thing. It's an incredibly killer, smart consultancy called Berg London. They just investigate things. Some of them in the way that you investigate things. But recently their investigations have taken the form of conceiving of themselves in China, manufacturing products which they themselves sell by direct mail. And if all of their products are very much as we described, they're retro-nostalgic jujitsu things."

Bill goes on to describe his favorite item. "I haven't gotten it yet, but I'm crazy to get one; it's called Little Printer. And Little Printer is a cube, it's smaller than a Rubik's cube. It's got an opening in the front that's the shape of an old fashioned analog television screen and the screen is a piece of thermal printer paper with a smiling happy face. Every morning their software prints you out your schedule for the day, the weather wherever you are, which of your friends are on which of your favorite social media, the top news stories. It prints it out the exact length of a bookmark on this little cheap thermal printer paper and you stick it into your daytimer or the book that you're reading. And at the bottom of it is the smiley face. The last thing it prints, which it leaves in the little printer, is the next day's smiley. So they're always the same. And it looks like an early 1960s Fisher Price toy. It's got two little legs and it's looking up at you. It's got a little USB cable. And you can set it to print any—"

Robert interjected: "Why does that—why does what he said," gesturing to me, "remind you of that toy?"

"Because it's taking you back," Bill replied. "It's not taking you back to the Victorians, it's taking you back to his birth year. Not to *mine*, but to *his*." He turned to me. "It's a like a toy from *your* childhood. The guys who are making it are your age and the people they are marketing to are your age and it comments on change and what's happening with print and what's happening with newspapers. It's really a very poignant little gizmo. And a great deal of thought went into getting it that way."

And then the other two members of the steampunk panel walked in: China Miéville and Dexter Palmer. My ambush interview was over, and we got down to serious business. I attempted both to politely contribute where it seemed appropriate and to keep my mouth shut where it didn't. I kinda succeeded. One of my great challenges as an interviewer (well, in life, really) is that I get excited and start talking too much. The good thing was that I didn't seem to have ticked anyone off.

Unfortunately, the panel itself seemed to go way over the audience's head. They were still grappling with lingering questions like, "Alright, these people up on stage are writing this really weird stuff—okay, we get that Margaret Atwood counts as important, but like, how does this fit?" Then the panel jumped all the way down the line to "What is steampunk?" It was like we just jumped the train off the track. I spent the rest of the weekend chatting with little old ladies who asked me to explain what steampunk is.

During those quite lovely conversations (the flippant caricature above intends flavor, not disrespect), I was distracted by a nagging thought. When the panel was talking about "what steampunk is," everyone answered mostly in terms of time and history—those pesky Victorians again. Everyone that is, except for Dexter Palmer. He didn't talk about the Victorian Age or the Industrial Revolution; he talked about cogs and transistors. I set out to find Dexter and figure out why he saw steampunk through a more timeless lens.

The Shock of the New: Dexter Palmer

> *There's nothing left that's miraculous anymore, and that's your loss for being born too late.*
>
> **— DEXTER PALMER**
>
> *The Dream of Perpetual Motion* (2009)

There's a lot of steampunk fiction out there, from the bad and the ugly to the wondrous and good. I had been in the midst of a streak of pulpy steampunk erotica, tawdry romance, formulaic adventure, and for lack of a better term, "gizmo porn"—that seems to exist merely to describe machinery in almost pornographic detail, down to the very last exquisite gear and burnished brass fitting.

Dexter Palmer jolted me out of this haze with his fascinating first novel, *The Dream of Perpetual Motion*. The book had made its way to the top of my reading list at least a half year before the Key West conference (Jeff Vandermeer speaks highly of it in *The Steampunk Bible*), but I'd been putting it off to churn through as much of the chaff as I could. One day, after finishing a particularly unfortunate book that (again) featured minor nobles cruising around Europe in airships, this time with a BDSM twist that was clearly intended to titillate but served only to flatten already one-dimensional characters, I decided I was ready to change gears (so to speak). I switched to the "good stuff" list and was very glad I did.

Dexter's book plays off Shakespeare's *The Tempest* and, to me at least, stirs in flavors of Roald Dahl's classic *Charlie and the Chocolate Factory* (a book not without a little steampunk flair in itself), and the haunting carnival feel of *Something Wicked This Way Comes*.

I caught up with Dexter after one of his panels and managed to find a half-hour window to tuck away for a chat.

"So," I got to the point, "some of what brings us here is taking that step beyond steampunk culture to start to ask the bigger questions, like: 'Why now?' and 'What's happening in our culture that makes all this resonate so much?'"

Dexter nodded politely, following me.

I mentioned that his book really stood out as I read through my piles of steampunk. "You engage directly right away with this whole idea of technology and progress and humanity," I said.

"If you look at a book like *The Difference Engine* and my book, my source materials were very different," he said. "I mean, I was looking at a lot of stuff from a kind of futurism that hails from around 1900, but also Fritz Lang movies. I also looked at early Disney cartoons that had this obsession with mechanical men and stuff like that. There's a Mickey Mouse cartoon called Mickey's Mechanical Man where he builds a mechanical man that he trains to box. Then there was a famous one with Donald Duck where he goes to the World's Fair and it's full of mechanical objects that do things that humans are supposed to do, and he has Donald Duck problems, right? Where he tries to use the device and it goes wrong, he gets angry and screams, stuff like that."

"And what is it that drew you to that material?" I asked. "What brought you to this place as an author in the early 21st century? Because you didn't, please correct me if I'm mistaken, come to this book with an obsession with steampunk or a fascination with gears. Something else caught in your mind's eye."

"I started writing this book in 1996," he answered. "I was doing research in graduate school for a paper on H.G. Wells and I came across this book in the stacks at Firestone Library—you know, one of those things where I found a book next to the book that I needed. It's the book that Jim [Gleick] mentions in this program—if you look at the two pages here..."

He opened the gorgeous event program and flipped through some of the images that had inspired his book. They'd also resonated deeply with Jim, who built the conference program guide around them.

"Yeah, I love those images." I responded. In fact, the instant I picked up my program right after registering, I knew I was in the right place just by glancing at its cover.

"I saw those," Dexter continued, "and I thought, there's something about the expressions on people's faces. It's not just the sort of oddball future that's half right and half wrong. I mean, you can often tell you're reading a science fiction novel because either the characters or the narrative voice are very excited about science, whereas the way that we actually deal with technology is that, you know, it becomes commonplace. You either take it for granted or you get bored by it."

"When I was a kid," he said, "I used to watch reruns of *Star Trek* on TV and Captain Kirk would pull out his communicator and then just talk to whichever other character he wanted. If you had told me that in 20 years I would have something like that, I would have thought you were crazy. Now I have a cell phone that does just that. It's not even a smart phone and I don't care. I'm not just bored so much by it as indifferent."

I knew this place. This is something Brian and I have been discussing throughout the Vintage Tomorrows project. "It's that point at which technology becomes ubiquitous and mundane, right?"

"Exactly," Dexter nodded. "One thing about these illustrations that I think is quite special—and that made me start thinking about this alternate world—is the complete indifference of the characters to what they're doing. If I were, for example, a character in a generic science fiction novel, whenever I made a phone call I would pull open my cell phone and then I would explain to somebody how it works for at least a whole paragraph, then I would get really excited and I would have this incredible sense of wonder."

Yep, just like the steampunk gadget porn I'd been reading. No wonder Dexter's novel woke me up.

"But in my own life, I only notice cell phones when they don't work; when I can't pick up a connection. So I liked the idea of writing a novel set in this alternate world, where characters in that world would take technology for granted and yet—and this is the advantage of using steampunk to write about technology—by changing the context and by changing the stuff that the machines are made out of, you can sort of combine that sense of the mundane with the shock of the new. You can show that it is kind of crazy that I can pull out my cell phone and talk to pretty much anybody if I have their digits, and yet at the same time I can do that and still have this indifference of how I respond to it. Whereas in straight science fiction it's hard to get both of those at once, do you see what I mean?"

"Absolutely," I responded. "I think what you're hitting on, too, is that as a fundamental tension that's going on in our culture right now, is that we have all of this stuff that has just fallen into the background..

"Yeah," Dexter said, "it's partly deliberate that there's a nostalgic past that the characters kind of share but isn't described in this book—it's a thing that happened, a shared collection of memories."

And that's a key part of the sophistication of *The Dream of Perpetual Motion*—it's set amid the spent wonder of a world both like and unlike our own. The world Dexter created is magical and mysterious in the way that our own might feel to a culture at a remove—perhaps a hundred years and one dimension to the left.

"When you look at the technology that's around you," I asked, "the way people are using it, maybe the way people are reading your books, talking about your books, do you see or feel anything kind of going on in our culture that might be a change? I ask because I wonder sometimes if there's some element in steampunk or in what people have been trying to do around the maker movement—in grappling with these physical things of the past—that is an effort to find some kind of anchor for the technological roller coaster we seem to be riding. Maybe trying to tie this down and make it somehow part of a human story?"

"Yeah," Dexter said, nodding thoughtfully, "and this goes back to not just the technological aspect of steampunk stuff but the visual aesthetic; a machine that's made out of gears is something we can visualize working. And because of that it inherently makes sense to us, so we're comfortable, at least in one way of thinking. Whereas something that has electronic parts in it, I mean it's a whole box, something magic happens in the box and then..."

I interjected: "It's ones and zeroes."

"Right," he responded, and then took an interesting step aside. "This might sound like a tangent but it isn't, some of the my favorite sort of science fiction films are like the big effects films from the 1980s, we're just like near the tail end of sort of big handmade models and stop motion, maybe the very beginning of using CG. And now when I see a movie, when I see an effect I just know they did it with computers. So at the same time it's like you can do anything, but there's no magic there."

"Is that maybe a question of artisanship or craftsmanship?" I suggested.

"Yeah, and this goes back to steampunk in a way—back to something I said earlier. The mechanical stuff is a way of allowing the interested layman who's not interested in abstractions, because learning electronics requires abstraction. It's a way to deliberately and concretely visualize a sense of magic that technology offers, which matters when you think about the shock of new things."

I had caught Dexter in a brief window of time between panels, but in that short conversation, he'd both highlighted and underlined the sense of magic and wonder I'd discovered in opening *The Dream of Perpetual Motion* for the first time. There was indeed something in the air here at the Key West Literary Seminar, and it didn't come from the cigar shops or margarita bars. These were the heady fumes of change. Steampunk sure plays into that, but it's clearly not the whole picture. After all, none of the authors here actually think of what they do as steampunk. That's one adjective among many. What we were discussing, really, was indeed the Literature of the Future.

ENDLESS SUMMER

What I learned in Key West is about the broader strokes of culture; about aesthetics, style, and perception. China's straight to the root approach that we saw back in Chapter 8 has a real place in this conversation because it reminds us that imagination is only half of the equation. The other half is action. I'm not sure that steampunk can meet China's high bar, but I do see it posing questions and presenting different perspectives. As to our culture as a whole, well, if there's anything being here in Key West has taught me, it's that we are starting to think differently about ourselves, our potential, and the world around us.

The first step is admitting you have a problem. The second (to bastardize the next of the traditional twelve) is recognizing that the power to overcome that problem exists. The third is to tap into that power. Technology sure ain't God. Neither is the human imagination (though in my personal opinion, it's pretty darn close). But put the two together and apply them to the real problems that plague our society? That's got potential.

I left Key West from the same quaint tarmac where I'd first arrived, walking out to my plane through the last bits of sunshine I expected to see in a long time. I was reeling from the feeling of epic change afoot—a new way of thinking that resonated with all my suspicions about steampunk. Our technological capabilities are truly beginning to dawn on us, and we're asking the right questions; tough questions. We're embracing possibility, but with open eyes and skeptical minds; a willingness to look backward as well as forward and to find answers where they live, not just where we expect them. As I basked in those last moments of winter sun, I felt a lot of hope for the future.

It turns out that my next step in this journey would bring more sun, and plenty more questions. I was on my way to Australia to investigate the flip side of the steampunk coin. In Key West, I spoke to writers about how we envision our world. I was going to Adelaide to talk to a tinker about how we actually build it.

The Answer's in Our Own Backyard

Of all the deep cultural and historical veins that run below steampunk, making and hacking have been around long before any of the subcultures came to be. Before there were "makers" there were teenagers who took a torch to dad's old Plymouth, transforming it into a magnificent hot rod. Before there were steampunks there were passionate and playful historical revisionists. And before we had hackers there were millennia of folks who just plain needed to get things done. Maybe they didn't have exactly what they needed, so they pulled a little from here and scavenged a little from there, but in the end with ingenuity and a little humor they made it work.

James travels to South Australia to meet Mark Thomson, a builder, maker, and a true original. Mark has been using comedy and history, high-tech and low-tech, to create art that tweaks our view of the past and challenges us to think differently about the future we might want to live in.

Empire and Rebellion in a Cup of Tea

James H. Carrott (Sydney, Australia)

> *"No, he said, "look, it's very simple... all I want... is a cup of tea. You are going to make one for me. Keep quiet and listen."*
>
> *And he sat. He told the Nutrimatic about India, he told it about China, he told it about Ceylon. He told it about broad leaves drying in the sun. He told it about Summer afternoons on the lawn. He told it about putting in the milk before the tea so it wouldn't get scalded. He even told it (briefly) about the history of the East India Company.*
>
> **— DOUGLAS ADAMS**
>
> *The Restaurant at the End of the Universe* (1980)

I arrived in Australia in a bit of a daze. I'd never traveled this far from home and the flight, while surprisingly comfortable, arrived late and I missed my connection. This meant I had a six hour layover in Sydney before my flight to Adelaide. With a little time on my hands I took a train to the harbor, and ended up wandering around the Royal Botanical Gardens. Located right next to the iconic Sydney Opera House, the gardens are a sprawling beauty, a fascinating bit of Victoriana in their own right.

Skyscrapers loomed in the distance but it felt wonderfully secluded. The trees we're alive with some of loudest birds I've ever heard. I stumbled upon a sculpture installation by Kimio Tsuchiya titled "Memory is Creation Without End." The sculpture is scattered across a wide lawn, with the city's glittering modern skyline in the background. It's an arresting piece—bits and pieces from demolished buildings. Tsuchiya intended that the work capture the circular connection of the past, present, and future—a wonderfully steampunk sentiment.

Part of the inspirational landscape of Kimio Tsuchiya's "Memory is Creation Without End," in Sydney (2012)

I had tea in the garden, scribbling away furiously at my notes for this book. As I sat amid layers of imperial past, my pot of tea seemed both perfect and problematic. An obsession with tea is one of my many quirks, and like my pocket watch, it's found intriguing harmony with the steampunk crowd. It's oddly reassuringly to see my own little contrarianisms echoed in a group of similarly-minded folks. You might think I'd be over that after a couple decades of counterculture. But nope, it's still strange. Not bad. More like a *Muppet Show* flavor of weird.

China Miéville essentially told me that "empire is empire." He thought that the steampunk impulse to embrace the stuff of Victorian Britain acted (wittingly or no) to support the contemporary neo-liberal imperial enterprise. And when it comes to the proliferation of uncritical steampunkery, I think he's right.

The stuff of rebellious style, culture, and fantasy invariably bite the hands that feed them. It is all too easy to appropriate cultural rebellion. While those rebellions do create change, they can easily end up the equivalent of electing a Democrat rather than a Republican to the White House. Did you fix the real problem? Nope. Still fat cat capitalists, just in cooler shoes.

The *real* American Revolution took a bullet to the head when George Washington and Alexander Hamilton rolled out federal troops to squash the Whiskey

Rebellion. The French Revolution is better remembered for the guillotine than its principles of liberty and equality. And I sure don't see a communist utopia in Russia or China. I'm not blasting the idea of revolution itself, just warning that it's a big gun, and big guns have a way of falling into big, powerful hands.

But I also have to believe in the power of the little rebellions. Maybe it's the part of me that jumps up and shouts "Yes!" when I hear Magpie Killjoy speak out in favor of "blind optimism." (My conversation with Magpie is in the *Historian's Notebook* (see Prologue).) Maybe it's just the cynical historian who looks back at the attempts at big and fast change we've made over the years and sees a spate of counter-revolution and regression.

In the larger scheme of things, my cup of tea is far from an act of resistance. But at the micro-level, *where change can actually start to happen*, it just might make a little bit of a difference. Maybe my quirks open an eye or two. Perhaps they cut through a stereotype, or make someone do a little double-take that shifts their perceptions a millimeter. To me, this little cup of tea underlines my sense of individuality and my sense of affinity with a culture I wasn't born into.

A couple white birds the size of small dogs wandered around the palms in front of me, pecking through woodchips with crazy long thin curved beaks, digging up some kind of insect. I sighed. It was time for me to go back to the airport. I looked at my "civilized" cup of tea. It was so simple and so complicated at the same time. Cultural patterns are historical, complex, and often hidden. We're poorly trained to see these connections and often miss that history is the stuff of power.

As I made my way to the train back to the Sydney airport, I wondered just how much steampunk really is trying to work out something new.

The Septic Tank: Overflowing with Ideas

When I arrived in Adelaide, my hotel provided further reassurance that my trip here was solidly tied to this project. A former government building, originally the South Australian treasury, it sat on a beautiful central square that featured a statue of Queen Victoria. The lap pool I swam in each morning was carved out of a former vault. It was solid past. Okay, fine: liquid-filled solid past. Upstairs, walls full of wooden plaques and chalk boards listed a century and a half of ministers and committees.

Heady with all this gorgeous past, I set off to meet Mark Thomson. Mark is a singularly impressive man. You wouldn't necessarily think so at a glance—his humble manner and friendly chattiness blend as easily into the small towns of the outback as they do into the halls of South Australia's parliament. Mark is entirely approachable and unassuming, as glad to chat cricket as to poke fun at the latest gust of political wind. But he's also a brilliant designer, artist, writer, and leader, whose occasionally self-deprecating wit belies an even deeper spark.

Mark would be the last one to tell you he's anything special, but that's part of the point: he is a paragon of the everyday bloke. His mission is nothing less than getting people to think, then putting tools in their hands and making them do something about it. His message is simple: *Think for yourself. Make stuff with your hands. Don't let anyone tell you that you can't. And* ***make it better.***

Mark is a connoisseur and unabashed advocate of "unacknowledged creativity." He finds it endlessly amusing that there are some kinds of creativity that are deemed "respectable" and others that are not. He's out to laugh this idea away with a tweak of the nose. At the same time he wants to create meaningful things. There's a hidden power to his approach; an intimate understanding of the power of laughter. Mark truly understands that wit and wisdom are strands of the same rope.

The Institute of Backyard Studies is Mark's "tink-tank." He calls it "The Septic Tank: Overflowing with Ideas." Like just about everything Mark does, it began with a joke. He told me the story as we climbed into his battered white pickup truck and drove from central Adelaide to his home in a nearby suburb.

"I started doing some work for a friend of mine," Mark started. "He was an academic that wanted to pull together a think tank called the *Institute for Social Research* . It sounded very serious." He gave me a sidelong glance and smiled. "So I did his business cards, letterhead, and envelopes."

"About 6 months later," he continued, "I was listening to the radio and heard my friend's name. The announcer on the radio said, 'And now here's so and so from the Institute for Social Research. He's an Australian expert on blah, blah, blah.' I thought. 'What? You're kidding me!'" Mark laughs with a sharp twinkle in his eye that speaks of intelligence.

"It was just because of the letterhead and the business cards," Mark shook his head. "Those things said he was an expert. So, I started musing"—Mark is forever musing—"if I was to start an institute with an official sounding name what would I call it? The Institute of Backyard Studies. That sounds pretty serious. But then to tell you the truth, I just put it out of my head."

"Then my book *Blokes and Sheds* came out," Mark explained. "I did a radio interview with a station in America. I was on the phone on hold waiting and the

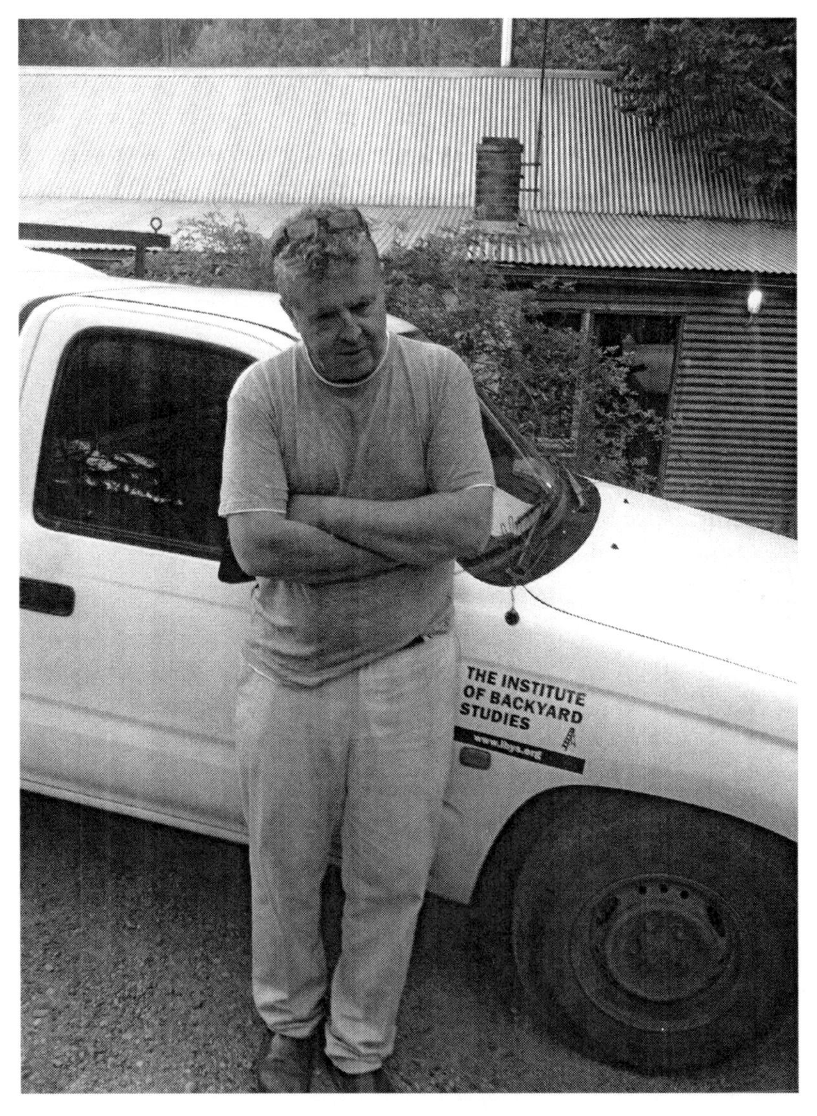

Mark Thomson (somewhat reluctantly) poses with the Institute of Backyard Studies' truck, my transportation through a world of inspiration. (February 2012)

producer came along and said, 'Where are you from Mark?' I said, 'I'm from Adelaide.' He said, 'No, no, what *organization* are you from?' He didn't quite understand what I was doing there or why I was a guest on the show. So I said, 'I'm from The Institute of Backyard Studies.'

"And the producer went, 'Right.'

"Sure enough a few seconds later, they introduced me as 'Mark Thomson from The Institute of Backyard Studies.' They didn't get it," Mark giggled. "So after that I did the letterhead and business card. It's taken off from there. We've strangely become something pretty serious. Tinkering has been culturally valued for a long time, but no one examined it seriously. That's changed over the last year or two. Now I get a lot of academics contacting me about all this stuff.

"It is also nice because the name *Institute of Backyard Studies* has a kind of joke in it," he said; you can sometimes hear the wink in Mark's voice. "A lot of the stuff that goes on in sheds happens beneath the official gaze. There's amazing things going on in the background—in the nooks and crannies—that is usually written off as being irrelevant. But I think it's really interesting and can reveal more than it seems at first. So 'Backyard Studies' is partly sort of hinting at that. It also has a domestic quality to it."

"It has a similar flavor in the States," I shared. "Homey, almost wholesome. I was part of a community radio station in Madison, Wisconsin, and we used to call ourselves 'Back Porch Radio.' It underlined who we were and our connection to our neighbors, who were both our listeners and fellow DJs."

"Right, exactly. That's it, you've got it!" Mark replied with a smile. "The back porch thing also implies a directness and intimacy. We've often thought of the idea of floating a political party called *The Australian Backyard Party*." He giggled again.

"We would only have our political meetings at a party in the backyard."

We both continued to laugh as the old white truck rattled along towards Mark's house. But we had stops to make before we got there.

Sorting Through the Rubbish

We passed through Mark's town on one of the occasional "hard rubbish" days. We stopped to scrounge through the piles of the stuff the council (local government) will only let people put out for trash on certain days. It was an odd mix of things, from broken computers and busted chairs to all sorts of bibs and bobs. Mark isn't alone in his treasure hunt. Some folks dig for salable scrap; while Mark's fellow artists and tinkers search for, well, different things.

"I am looking for stuff that doesn't necessarily have a strict purpose," Mark explained. "I'm looking for things that are adaptable or have qualities that I think have potential uses. Like for instance television aerials (antennas). TV aerials are

A selection of Institute bumper stickers. The axiom at bottom now adorns the back end of my little red Honda Fit at home in Seattle, and provided a fitting title for this chapter to boot (puns intended). (photo courtesy of the Institute of Backyard Studies)

made out of aluminum. Now that people are getting newer TVs they are getting rid of their old ones. Now there's a lot of embedded energy in aluminum and it's easy to form into nice shapes. Also aerials are really hard to make so I collect them and put them aside."

Mark has a friend that does mosaic work out of found glass and ceramics. He keeps an eye out for stuff for her and she does the same in return. Relationships keep the community together.

They're not recycling, they're doing one better: they are repurposing. Mark and his friends place the highest value on real use—not just sticking old computer chips onto jewelry. They look to leverage and use the functional parts of what they have found.

In steampunk terms, this would be the difference between a clever, self-made costume and slapping a pair of mass-produced goggles onto a store-bought top hat. The difference is depth of thought—actual thinking—a real creativity that lives beneath the too-easy surface. This type of crafting and hacking is usually done with a laugh in your belly and a smile on your lips.

I've learned that Mark has three main types of facial expressions: boyish enthusiasm, mischievousness, and deep thought. He wears the last as he continues:

"When I try to have a discussion with the people who make a hair clip out of an interesting piece of their computer it never goes well. They usually get offended. I think they could do much more than just recycling. There should be more substance."

As you know, I think that image can be substance. But cosmetic steampunking does seem to fall flat beneath Mark's reminder that *you could do so much more.* Mark is no steampunk. His work holds an interesting mirror up to steampunk makers and modders. However, I didn't come all the way to Australia to stay obsessed with steampunk. I was here to explore bigger things, things that have been going on in Australian culture for centuries. It all ties together at the heart of our shared impulse to create.

"James, come give me a hand," Mark called out from the curb.

I grabbed an armload of anodized aluminum door runners and old fence posts and loaded them in the truck. Starting for home he mused, "I'm thinking of twining them together, sandblasting them, then powder-coating it all into a gigantic interesting compound beam for the new shed."

A Bloke Without a Shed

A "shed" in Australia is a backyard workshop, usually a corrugated metal outbuilding. In my week there, I became convinced that most of Australia is built whole-hog out of corrugated metal. This makes sense, as Mark later explained, because it's inexpensive, durable, and lightweight. Where an everyday American guy might retire to the garage to tinker, an Aussie bloke's realm is the shed.

When I visited, Mark's shed was down for the count—he was a bloke without a shed. The old one had been claimed by termites. Now he got to plan a new one. He just needed to jump a few pesky zoning hurdles.

Without his shed, the contents of the previous structure were scattered across his yard. The scene was a battleground of chaos and order—piles of scrap metal, collections of small engines and a vast array of metal and wood minutia. Pulling our hard rubbish haul from the back of the truck, Mark and I set out into the piles to find the proper place to "store" the new items.

We came across a pile of beautifully-cut metal fence posts that Mark identified as being cast-offs from a 1950s industrial process.

"Look at that. Beautiful ones. Pretty high quality at that. This is something that I got very recently," he said proudly. "These are from a house that was being de-

The Artist Formerly Known as Mark's Shed

molished up the road and the guy said, 'Sure you can have them.' I haven't worked out yet what to do with them, but I know a whole bunch of people who also collect things like this, and every one of them say, 'Yeah! They're good, they're good! Dunno what for, but they're good!'" Mark laughs at his own jokes.

Mark's reputation as a tinker paragon has taken on a life of its own. "People have started to give me all sorts of interesting old mechanisms like these," he said, holding up a beautiful bit of vintage machine something-or-other.

"What is it?" I ask.

"Old farm machinery. I really like the aesthetic of farm machinery. It's really tough but rusty. It's really *hard*, you know?" He paused thoughtfully. "I really need my new shed. It's frustrating not to have all my stuff laid out around me. I'm not able to draw on all those permutations, all the possibilites," he motions to the piles. "For instance all that is fairly good quality or interesting timber." He points at a massive stack of wood. "And we've just got those lovely pieces from the hard rubbish." He points back to the new treasure. "And, oh! That is off a piano." He continues to point around to gorgeous bits of wood and metal tucked under eaves and tarps, piled up against and around his house.

The lack of a shed made Mark's house look turned inside-out. We chatted a little about this as we meandered through what sometimes seemed more salvage yard than backyard.

"All of this is just so easily misunderstood," Mark said still serious and thoughtful. "It's badly assessed. Do you know the American TV program *Hoarders*?

"Yeah I do."

"It's terrible," he said with a rare frown. "They vilify people who just collect stuff."

"I know lots and lots of people who are good users of that stuff. I'm a fair user. A whole a lot of this stuff, I've just sent to the U.S. as part of an exhibition. The entire art exhibit was made from these piles. This is all just raw materials," he surveyed the contents of his demolished shed. "I recognize the capacity within these things. For instance these," he picked up two sets of venetian blinds. "We call them venetian blinds. What do you call them?"

"Venetian blinds." I smiled.

"They are exquisite things in their own way," He ran his fingers over the blinds. "They are very very thin aluminum and they have a special interesting quality. You can actually cut them with scissors they are so thin. They are metal but they are also almost like leaves. They are beautiful."

Seen through Mark's eyes, old venetian blinds are a treasure. They are wildly adaptable for use in a dozen gorgeous creations. That speaks directly to Mark's character—he sees potential—not just the objects themselves, but what they could become.

Mark kept rummaging. "Some of this is just interesting historical stuff. That's from a Harvard Trainer, a World War II airplane. See that metal? It's actually dualuminum, not aluminum. Those specific pieces were from a plane that crashed in Central New South Wales. A whole lot of those planes ended up here as crop dusters."

Everything in Mark's un-shed has a story. I was stunned by how much history is tangled throughout his yard. And that's saying something. I've poked around in the archives of the American Philosophical Society, prowled the closed stacks of The Newberry Library, and explored storage rooms at the Royal Ontario Museum. Mark's yard ranks right up there with all of them. There was old furniture, a crate of steamer inserts that look like polished flying saucer parts, a pile of vacuum cleaner tubing, and a collection of old cricket bats. I paused by the bats, and Mark noticed me stop.

I'd had a cricket bat before—a present from my father, who was always trying to entice me into one sport or another. (cricket might actually have worked. I love the quirky character of the sport, much in the way that I later came to love baseball.) I had sold the bat in an ill-advised fit of stuff-reduction during my move from Massachusetts to Seattle, when I left my dissertation behind to begin a career in games and toys. I've always regretted the loss of that bat.

Mark gave me one of his.

A Wall of Hammers

Stepping inside Mark's house reminded me just how hot and sweaty I'd become. It was February, but in Australia that was summer. I was parched and sunburned.

Seeing my state Mark offered, "Lemon lime and bitters?"

Always say yes to lemon lime and bitters.

Amazingly refreshing drinks in hand, we stepped into the living room. Mark's house was a bit like entering the eye of a hurricane after wandering the shed-less yard. It was relatively calm and serene. Inside the house I saw how Mark turned the chaos of the yard into brilliance.

"I have been trying to build a Wisdom Receiver," Mark said, gesturing at a table full of fantastical oddments that already included an "Ignorance Amplifier." I guess it called out for balance. "These are various parts of it and the rest are off being made. That's the trouble working with a network of conjurers," he said, referring to his sometime collaborators. "They are conjurers and they often take a long time to do anything."

He continued to show me around the informal museum of brilliant oddity. "This was an earlier model of Wikipedia in a box." The simple looking box had a label on it that read: "Pocket Pundit."

Then I froze, enthralled by an entire wall covered with hammers. "You've got a lot of hammers," I said finally.

"Yeah, a bit of hammers." Mark is a master of understatement. "My wall of hammers does disturb people."

"Are you saving those up for something?"

"Well, I just sort of started collecting them and then people started sending them to me," he shrugged. "I like a good hammer. It's a good tool."

I'm reminded of Cherie Priest's insight (Chapter 4). No matter how fancy our tools become, there will always be a need for the humble hammer, a solution for everything from unpounded nails to zombie apocalypse.

"Let me show you the kitchen," Mark continued the tour.

A bit of "a bit of hammers." Seriously, this is just one part of one wall. My phone's camera is too limited a tool for this majesty.

The kitchen wall was decorated with a dazzling display of wire.

"Were you doing a show on wire?" I asked.

"I was doing a whole book and a film on wire," he answered. "So I started collecting all these little things here. Things like a bush made of clothes pins."

"That thing is for killing snakes." He pointed to the wall. "The guy who gave it to me said that when he was in the army in the 1960s every army vehicle had one of those in it.

"It looks handmade," I said stepping closer to the snake killer.

"Oh yeah yeah," Mark replied as if it was completely natural that a government-issued snake killing device would be handmade. "The thing will flex and hit the ground, it turns flush to the ground then you go like that," he flicked his wrist. "It's the best possible thing to kill a snake with."

"What's it made of?" I was fascinated.

"It's just fencing wire," he replied casually. "I'm going to turn all of this into a traveling show."

"Why wire?" I asked, pulling myself away from the wall.

"It's about potential capacity," Mark said while scanning the wall of wire. "Wire has the capacity to be shaped into anything at all. That is the beautiful thing about it; the possibility. It also increases our human capacity. There is an interesting relationship about capacity in that way. Some things have the potential to increase our human capacity. They allow us to project our own capacity through working with our hands. That's what wire can do. It lights up a specific part of our brains."

Henry Hoke, Tinker of a Deeper Truth

Mark's 2007 *Henry Hoke's Guide to the Misguided* is a deeply funny book that chronicles the life and work of the enigmatic Henry Hoke. Its subtitle alone: "An Inventor's Thwarted Genius—A Speculative History" plants it solidly in my line of inquiry.

You could travel all around the world and never meet a man like Henry Hoke. There's also one of him in every town in the English-speaking world (probably the rest of it too, I just can't speak for them). Hoke was an innovator, a creator, an enigma, and perhaps (but only perhaps) a figment of Mark Thomson's robust imagination.

Henry Hoke is not steampunk. That said, *he's all about what it means to mess with the stuff of the past.* Hoke, or the idea of Hoke, plays with the cultural bits and bobs that make up the stuff of our everyday lives. The stuff of Hoke's world can turn your mind sideways.

Henry Hoke invented the Learning Curve. Though it's become a metaphor for how quickly people learn a new skill or digest new information, it began as a meter-high metal curve mounted on a heavy wooden base that sat on Hoke's shop floor.

When an unwary apprentice or visitor walked into it, the curve delivered a whollop to the inattentive victim's kneecap or shins. Folks quickly learned not to walk into the big metal thing on the floor. The Learning Curve imparts practical wisdom: watch your step.

This wit is central to Mark's work. The ever-developing mythos of Henry Hoke is based on it. It's not enough just to be clever or funny, but to *have your wits about you*. The Learning Curve is a great illustration of this principle. It's funny on a couple levels. Starting with a Three Stooges-esque smack on the shins delivers a healthy whollop of clever wordplay, leaving you a clear message to pay attention to the world around you: *use your wits*.

Henry Hoke has a long list of witty and amazing inventions. There's "Hoke's Waterproof Tap" which really helped in South Australia's long droughts. And who could live without "Hoke's Wooden Magnet" or "The Headless Nail"? Other popular items included the "Glass Hammer" and "Hoke's Load of Balls" (though the function and purpose of the latter remain a mystery to most).

Silas Hoke, Henry's pharmacist father, was known for his bottled "Willing's Suspension of Disbelief" which Mark chronicles as "the patent medicine that kickstarted the Hoke Family dynasty." And who could forget "Dr. Silas Hoke's Dehydrated Water Pills." And Mark says he doesn't do steampunk...

"Alright," I said, "How did you first encounter Henry Hoke? How did all that come about?"

"Well," Mark leaned back in serious thought. Well, it could have been serious thought. He might have been poking fun at me. Hard to tell, but it didn't really matter. It'd be smart and funny either way. "Henry's about more than what we've talked about so far on your visit. But it will take a little explaining."

"I started off working in the unions," he continued, "and I would hear these stories about people telling the new kid in the shop: 'Go and get the box of spark plug sparks.' These were tricks that the workers played on apprentices to test their skepticism. They wanted to see if the apprentice was listening. It also initiated the newcomers into the tribe, the union. If the apprentice really thought about it they'd get that there was no such thing as a box of 'spark plug sparks.'"

"Dr. Silas Hoke's Dehydrated Water Pills" (photo courtesy of Mark Thomson)

"So it's like a kind of hazing?" I asked.

"Not really," Mark replied. "It wasn't about violence. I typically think of hazing as violent. I've spoken to quite a few older tradesmen about this stuff, the stuff that happened to them. These are people who'd had numerous apprentices themselves. What's interesting was that it was actually about questioning authority. They were trying to get the apprentices to really listen to what the boss was saying. Is it bullshit or not? They wanted the apprentices to think for themselves—to ask: 'Should I take everything literally?'"

"Henry Hoke came out when I was doing the *Rare Trades* book on artisans and their unique skills," Mark explained. "I came across these tricks time and time again. So I starting turning them into objects. I'd make the can of striped paint or a container of left handed screw drivers. It was good fun. A little while after I started

"Spark Plug Sparks" (photo courtesy of Mark Thomson)

playing around with these I was talking to a friend of mine in Sydney who's in publishing. I asked 'What am I gonna do with these? People really like these things.' And he said: 'Well, why don't you just find something that ties them together. Think of them as a fictional story of some kind.' And I thought, 'I can do that!'

"So I created Henry Hoke, the guy who invented all these things. I'd already been doing other books about inventors and people who were innovators so it happened very quickly. Even his name Henry Hoke is a trick. If it's a fake, it's a kind of hoax." Mark's humor only gets funnier with repetition, and he takes such glee in this wordplay you just can't help but laugh with him.

"It grew from that," he went on. "I kept thinking, 'Well, he was *somebody*.' He had his own company: Hoke's Tool company. He lived in one of those towns that are all over South Australia, and his family had been there for a while so they had their own town: Hoke's Bluff."

"Hoke's Bluff," I repeated, unable to stifle a grin.

"Yes, Hokes' Bluff," Mark replied. "Bad jokes and bad jokes and bad jokes..." He continued with a wry smile. "The pleasure is in the language. It's an old-fashioned type of humor. It's almost a kind of steampunk humor; clean humor, gentle and clever."

"Well," I said, "It does turn things on their head. Tweaking them..."

"Yeah, it's a gentle little flip. But I do think it has a use," Mark said humbly. "The literalness of our world now is in itself kind of absurd."

"What do you mean by that?" I was hooked.

"People are so serious and literal about things that humor really only lives in this very defined little area. This is the kind of humor that can creep into everyday life." Mark thought for a second then added, "Maybe it's not so important..."

"No," I spoke up. "It's a really big deal. It's very important. It's one of the three big things from our research for this project. People want humor, history, and humanity from their technology. Having a sense of humor is really important."

"I like the history part as well," Mark moved on. "What I like about Henry's technology is that it contains mysteries beyond the grasp of the 21st century mind. But they are ideas that were perfectly reasonable and logical in the past. They don't work for us today because we think we understand everything else."

Now this is some meta-steampunk thinking. I held back from using the word, of course, but what a brilliant insight into the mysteries of the past—and, perhaps more significantly—a sharp observation about contemporary culture. We *think differently* now.

"Are you saying that the way we think has changed our reality?" I asked. "You're saying that Henry's inventions worked until we changed our minds?"

"Yeah, our own logic has confounded us." Mark stated with an authoritative wink to his voice. "There are things that are simply beyond the conception of the 21st century mind."

"Henry Hoke makes people think about how we've changed the world by thinking differently?" My mind was blown with an almost audible *pop*. You might think I'd be used to this kind of thing by now, but I'm really, really not.

"Henry's Random Excuse Generator is an interesting notion," he added, "Because it runs on refined bull dust it's able to successfully transfer matter into ideas and from there into language. So that in itself is somewhere we just won't go in the 21st century. Directly turning matter into ideas. It seems like you can't get there from here."

"That reminds me of something William Gibson talked about when I was in Florida," I said. "He thinks that when people generations from now look back on us they will think it's quaint that we distinguished ideas from objects, reality from virtual reality. To them it will all be the same thing."

It has to do with the idea of the digital world. Matter to ideas. Ideas to matter. Just how much does matter, well, *matter*?

"Hoke's Random Excuse Generator" (photo courtesy of Mark Thomson)

The Hand-Brain Link

We met for breakfast the next morning, our last chance to chat before I hopped my plane to New Zealand. (That adventure is chronicled in the *Historian's Notebook* (see Prologue).)

Mark put on his thinking face. He started, "I went for a walk early this morning. I was thinking about what you were saying about what William Gibson said: that people at some point in the future will wonder why we ever distinguished between virtual and real. That's what you were saying right?"

"Yeah," I replied.

"I don't know about that," Mark said. "I think our genetic history as humans is that we are tool makers. All the stuff I do is based on that. The hand-brain link is profoundly hard-wired into us. It lights up a part of the brain that is hugely satisfying. I think people yearn for the tangible. We've been making things for a long time.

"Building things, physical things, is a very serious part of human history," he continued. "Making things well means going beyond the surface or patina—it's about embedding some of yourself, the best part of yourself ideally, into that object. That transfer is a deeply complex process: it's how people make things that have a strong meaning attached to them—because those things have a part of their maker in them. Which is why I enjoy having fun with it," he said seriously. "But you have to understand when you mess around with it, you are messing around at your own risk."

"When you are messing around with history," I added. "You're not playing with Nerf balls. You're playing with golf balls. You throw them things around, and you're going to break windows, you're going to put somebody's eyes out."

"I think I'd feel better if steampunks were all making things with their hands." He finished, "If people are good with their hands they are usually good with their minds as well."

"I think we can get to a balance," I replied. There are people who dig deep with their steampunk—it's well-researched, clever, and powerful." I paused to see how Mark would respond. He said nothing. He was listening.

"That was my initial reaction to Henry Hoke," I continued. "I thought it was some of the smartest steampunk stuff I've ever seen. It taps into the deep veins of culture and history that we've been studying with this project. Henry Hoke is all a little off, it's using history and wit. It's like you're building cathedrals but you're building them upside down."

Henry's mother seems to have understood the hand-brain link quite well. From Henry Hoke's Guide to the Misguided: "This is believed to be Beryl Hoke's Hammer of Righteousness, clearly showing signs of frequent use." (photo courtesy of Mark Thomson)

"I think steampunk will migrate," Mark said. "You have all these people and they have web sites and they think that's pretty bloody advanced. But they never think about the permutations. So that's where I think some of steampunk will end up going. It looks like they are getting more firmly rooted in making and building. I find great hope in these Maker Faires (we visit Maker Faire in Chapter 11). They seem to get the interesting potential permutations." Mark seemed hopeful. "That's what America has always done well is finding interesting possible recombination of things." We both grow quiet as my visit drew to an end.

Mark gave me wisdom I'll carry with me for the rest of my life. Things that are worth doing are worth doing right. Things that are worth punking are worth punking right. And if you're going to mess about with something, do something real and meaningful with it. Don't screw it up. *Make it better.*

As I checked my newspaper-wrapped cricket bat at the Sydney airport, I couldn't help but think about how my life had changed in the past few decades. Yet the universe has a funny way of bringing things back around. I have a cricket bat again, but one with a warm story and rich history. It still needs a good cleaning and a generous application of linseed oil, but it speaks of past and promise rather than expectation and disappointment.

Sitting at the airport, firmly planted in the slip stream of travel, I found myself changed. Somewhat unexpectedly, I was leaving Australia a different person

than when I arrived. I've learned something here on the other side
I hadn't truly internalized in the previous 39 years of my life: Any
ble. Not just by imagining it but by actually doing it. It's not enough to live a life of
the mind. Dreaming doesn't suffice. You have to *build* your dream. And even that's
not enough, because what you build *better be good.*

Makers and Burners

James, Brian, Byrd, and the documentary crew head to San Mateo, California to attend the 2012 Bay Area Maker Faire. Amid a dizzying collection of makers, hackers, builders, and steampunks, James chats with two significant figures in his journey from the desert of Burning Man to the parlors of steampunk. Toward the end, he stumbles upon an unexpected space-time portal from which steampunk might just have sprung.

Cultural Capital

James H. Carrott (San Mateo, CA)

> *Expect the unexpected, Edie was told by a sour veteran sergeant in Burma, and the expected will walk up to you and blow your expectations out through the back of your head. Expect the expected, just don't forget the rest.*
>
> — **NICK HARKAWAY**
>
> *Angelmaker* (2012)

MAKER FAIRE STARTED in 2006, put on by the folks at *MAKE* magazine (they also happen to be our publisher—more synergy!) to, as they put it, "celebrate arts, crafts, engineering, science projects and the Do-It-Yourself (DIY) mindset." The San Francisco Bay Area Maker Faire is where it all started.

It's also where Burning Man started (the Bay Area, not Maker Faire), and where Ken Kesey and the Merry Pranksters launched the trip that echoed around the world. The Bay birthed the personal computer, too, and transformed the world through subsequent waves of technological wizardry. You've followed my wanderings long enough now to suspect that I see more pattern than coincidence in these facts. Steampunk is one thread among many, I think, as I trundle myself onto the hotel shuttle that carries me through the unfamiliar suburban byways of Silicon Valley. It's interesting, though, I feel this place in my gut. I've learned to trust my intuition about such things. I'm actually expecting something spectacular. There's a piece of the puzzle to be found here in San Mateo. And it might just be the one that snaps the whole thing into focus.

When I began this journey, I had no idea how deeply personal this book would become. That changed on a rainy Portland afternoon maybe half a year before, when Brian and I sat down to coffee at Powell's City of Books (another contender for "center of the universe" in this bibliophile's humble opinion) and he brushed away my chunky manuscript with a mischievous smirk.

I'd learned enough about working with Brian by then to know that he intended some sort of compliment in this dismissal. He's as socially nimble as he is insightful, so I stifled what I might have otherwise seen as a slight and waited for him to flip over the next card.

"I read *The Electric Kool-Aid Acid Test,*" he began, holding my clear enthusiasm at bay with another wave of his hand. "And I think you're right that there's something there." I grinned, glad that the deeper parallels I'd begun to draw in my own head were resonating in his. I wanted to jump in, but he still had that look—the one that says he hasn't gotten to the punchline yet.

He reached to pat the now lonely pile of paper while he continued, smirk extending into a broader smile, "This is our launching point, but I think we can do something better." My mind raced around for purchase—we'd been talking for a while about how, even with the film, there was something missing in our approach. Our topic was clearly a whole greater than the sum of its parts, but our book didn't carry that yet. Something finally clicked as I met Brian's knowing gaze.

"We're going to be Tom Wolfe," we said in unison. From that moment, everything changed. Vintage Tomorrows became a *story*. Our subjects became characters and *we* became characters. All of a sudden we were free to explore our subject in entirely new ways. Inspired by Wolfe's approach to creative nonfiction (and maybe a wee bit by Hunter S. Thompson on my part), we set out to do something truly different.

Why am I telling you all this now, closer to the end of my journey? Why here at Maker Faire? Because what happened that weekend tied a knot for me. It brought me back full-circle, changing my life yet again in the process. That missing piece turned out to be a piece of myself as much as of steampunk, or of the broader cultural patterns I followed to get here.

Friday: The Infrastructure of Inspiration

> *Let us think the unthinkable, let us do the undoable, let us prepare to grapple with the ineffable itself, and see if we may not eff it after all.*
>
> **— DOUGLAS ADAMS**
>
> *Dirk Gently's Holistic Detective Agency* (1987)

Arriving at the San Mateo County Event Center the following morning was a little underwhelming at first. My cab drove into a massive weed-dotted parking lot and drove up to what looked like the back of a concert building. Following my usual pattern, I was there early, a day before the big show began. Brian, Byrd, and the film crew would arrive the next day. I'd proven my early bird point catching Margaret Atwood in Key West (she's clearly no worm, though we did discuss one;

a super-brained megalomaniac worm from outer space, but a worm nonetheless, so I'll brook no snarky comments about the trite idiom). There's a downside in arriving early, though. The parking lot was nearly empty and it wasn't quite obvious where I was supposed to go.

So I did what I do in such cases and set off wandering, in search of some center of gravity, or at least someone in an official looking t-shirt. Beyond the fence, however, it became clear that I needn't have worried. As I got closer the awesomeness of the event came into view. It was a glorious chaos. A spectacular spectacle coming together before my eyes.

It was a sunny afternoon and I settled myself into the midst of mad preparations. All around me people were building elaborate booths, dragging in massive crates of gear and fighting against the gods of misfortune to make sure that everything was set up for the next morning. I spent most of my time in the steampunk area, since *Vintage Tomorrows* would have a "steampunk confessional booth" set up along the far side of a tiny clearing. It turned out to be the perfect place to watch Maker Faire take shape.

Josh Tanenbaum was to my right. He'd been on the fair grounds for days, gathering supplies, putting together tents, weighing things down with sandbags, coating picnic tables with chalkboard paint, and basically doing whatever was needed to make everything work. Josh is a bit of a saint that way.

Josh was pulling together the tables for the Steampunk Academy—my comrades in arms at the Faire. The Academy crew can build just about anything from loose scrap and maybe a trip to the local hardware store. Typical makers that way. Pretty much everyone here is like that. If they're not, they're on their way—learning and testing. That's the point. Well, that, and inspiring similar madness in others. Infectious inspiration.

I also met Kate Compton at the booth, whose gorgeously laser-cut business card introduces her as "Exceptional Genius; *Maker of Many Interesting Things...* 'Don't Worry; It's Supposed To Do That.'" She had a hacked Kinect for Xbox 360 with two screens and a dollhouse full of Victorian parlor tricks. It was fascinating to see Kinect, a product I'd spent so much time working on while I was with Microsoft, being put to a use I'd never imagined. I've come to think of these moments of "wow, I never thought about it *that* way before" as the raw stuff of inspiration. Spend enough time around places like Maker Faire and you start to recog-

Steampunkery by Josh Tanenbaum at the "Steampunk Academy" (photo courtesy of Alan Winston)

nize the early symptoms of personal paradigm shift: a particular sort of pause matched with a momentary far-away look, generally followed by a slightly upturned lip, narrow but twinkling eyes, and often either frantic scribbling or incessant babbling (in my case, both).

Kinect didn't end up being the perfect outdoor toy, but trying things out and seeing how they work in new contexts is at least half the fun. It's also an essential part of what connects Maker Faire with Burning Man. There's a kind of mad science in action that pervades both places—a desire to push the bounds of the possible and "make it work," no matter the circumstance. For more than a few folks here, Maker Faire is a kind of technical dress-rehearsal—the "here's how it works" before they bring it out to the Playa and set it ablaze (literally, figuratively, or both) at Burning Man. But the connection between Burning Man and the Maker Faire goes far beyond that—it's about the art of the possible.

Folks came to Maker Faire with some plans, bits and pieces of machinery, and contraptions in various stages of development. It seemed most had a weird and wide array of tools, and big rubbermaid tubs of "raw materials." It seemed impossible that this glorious chaos would somehow form itself into one of the largest maker happenings in the world in less than twelve hours, but I knew it would.

That's what the Faire is all about. Doing it yourself is a messy process. It generally involves breaking before the making and while plans are nice and useful things, they tend not to stand up altogether well in the face of reality. Who was it who said that few strategies actually survive an encounter with the enemy?

Doctor Who geek that I am, I was blown away (figuratively) by Cory and Hannah Soto's giant Dalek. Maker Faire, San Mateo (2012)

Expect the unexpected. Where did I stumble across that selfsame Dalek again? The Playa. Even in this hostile environment, they were still kind enough not to exterminate me. Dalek by Cory and Hannah Soto. (photo courtesy of Andy Pischalnikoff)

Maker Faire is an eclectic mix of booths, vendors, and educational sessions. You can learn to make pretty much anything on the planet here. If you can think it up, there's a solid chance you can find parts, skills, collaborators, or even a kit for it at Maker Faire. It's a giant hands-on enablers' party. The bane of tidy housemates the world over.

Josh and I took a lunch break to wander the fairgrounds. As we walked, I overheard things you just plain don't hear in everyday life, yet here it feels completely unremarkable. Maker Faire is one of those places—the ones in which I find myself more at home each passing day.

"Dammit, I should have brought my Burning Man rebar," grumbled one frustrated maker. "And of all the things I brought... no crowbar."

"I've got a crowbar," someone nearby happily spoke up.

"So far, I haven't heard anyone ask for a tool someone that someone else doesn't have," I said to Josh.

He nodded with a smile, but followed with a sly grin. "I try to keep it simple," he replied. "All you *really* need is duct tape and WD-40."

"I get duct tape, but why WD-40?" I ask, watching an army of blue-shirted volunteers pulling computer gear from the back of a van.

"Well," he begins with a dramatic pause. "There are basically just two kinds of problems in the world, right? Something moves when it shouldn't move—duct tape. Something doesn't move when it should—WD-40."

"Sensible."

On our walk I saw a giant umbrella tree, whose various appendages sprinkled rain and spit fire. That thing seriously rocked once it gunned up for the show. We also passed a full-scale geodesic dome assembled from flat pieces that had been cut into a couple simple shapes. It looked like one of those wooden dinosaur models you can buy in museum gift shops and high end toy stores, only simpler and much bigger. The entire structure flattened down to a few small basic two-dimensional forms that could be assembled without tools. Next to it sat a giant robot puppet face that was run by a series of controls some thirty yards away. The range of expressions it produced rivaled the complexity of the human face itself. Another inspired contrast—beautifully nuanced complexity next to pragmatic simplicity. But both the epitome of elegance.

We walked across the courtyard past a Cylon centurion from the 1970s *Battlestar Galactica* (an imagined world near and dear to my heart), past what looked like half a small car chassis set up to spin around like a windmill on a mini golf course. That turned out to be a "Colonial Viper simulator." I missed out on that one, but smiled at the memory of a childhood dream fulfilled some years back when I visited the set of *Battlestar*'s mid-2000s remake (WizKids, the company I was working for at the time was making a collectible card game based on the show) and got to sit in the cockpit of a *real* Colonial Viper. The astounding attention to detail I saw on that stage set stayed with me ever since. Every book and slip of paper in

Commander Adama's office had its corners clipped in the show's iconic octagonal style. I thought of the Neverwas Haul then, and the incredible significance of the little things—how the framed period-style photographs and upholstered parlor furniture kept me immersed in that imaginary past.

I'm reminded again of *The Hitchhikers Guide to the Galaxy* (that happens): "Time is an illusion. Lunchtime doubly so." Well, imagination is layered. Illusion doubly-so. Because the right peek behind the curtain—on a television set, at Maker Faire, at Burning Man—can fuel and inspire. This is one of those open questions, the contradictions that steampunk embraces—the garage punk's wrench and the stage illusionist's steam both enable the imagination. I was always more intrigued by the sharp-witted stranded balloonist pulling on knobs and twisting dials than I ever was by "Oz the Great and Powerful." The truer magic is woven into the making. Why truer? Because *you can do it too.*

When we got back to the booth Josh started taking apart power drills. You know, 'cause inside power drills, *that's* where magic happens. "It's amazing what sludge comes out of these things." He holds up a gutted drill. "They put the pieces together and then just fill them full of this thick grease." Josh scoops and scrapes for a bit then remarks more to himself than me. "I'm glad I didn't have the kids do this part themselves."

A big component of Maker Faire is teaching kids to make. Really it's about teaching everyone to make, but there's a heavy focus on getting kids into the mindset that they can build anything they can imagine. But Maker Faire also embraces the messiness and sometimes dangerous part of making. If you are going to weld or use a soldering iron, it's going to be hot and you have to be careful. Once you start actually building things, there's danger involved. We've already explored the ways in which the same is true of history.

Saturday: Mad Science and the Safety Switch Problem

Saturday came and the rest of the *Vintage Tomorrows* crew arrived. Byrd and I had an on-stage commitment. Byrd showed an early clip of the documentary and I talked about what I'd learned from my research. It was a bit preview, a bit presentation, and—having been thrown a curve ball by some technical difficulties—just a tiny bit vaudeville.

Our ostensible reason for being here; James H. Carrott and Byrd McDonald present Vintage Tomorrows during their panel at Maker Faire Bay Area 2012. (photo courtesy of Shawn Sundby)

With the on-stage bit behind me and my nerves settled (I have no fear of stages, but was unused to television cameras; in the end I just thought about Mimi and B watching me back home—and you can too (*http://bit.ly/VTProgram*)—and got through it just fine), I was ready to go about the rest of my business at Maker Faire.

And that business was dear to my heart, because it connected me back around to where I began this journey. Even before Brian and I sat down to talk past and future at Pike Place Market (Chapter 1), there was the Neverwas Haul (Chapter 2). Nothing before or since has captured my imagination like that magical driving house. A framed picture of the Haul sits on my writing desk at home, and its pipes, staircases, and balconies appear in just about every sketch or painting I've done in the past 5 years. The darn thing's gone and made an indelible imprint on my soul.

I'd connected briefly with one of its creators over Facebook and had "set up" (more like planned to plan) an interview. Unfortunately they wouldn't be able to bring the Neverwas Haul to the Faire; it was being rebuilt for its fifth anniversary return to the Playa. I was heartbroken about that until, well, we'll get there in due time. What I discovered right away is significant enough to start us off.

Meeting Kimric Smythe was... an experience. Kimric is a seriously interesting guy. Sometimes you say that about people and, yeah, everyone's interesting in one way or another. But Kimric? He's a beast of a different stripe. Kimric is the owner and proprietor of Smythe's Accordion Center in Oakland, CA. That's his day job. He and a small crew repair, restore, and sell accordions. This kind of thing requires a vast supply of obscure parts, and a sharp eye for finding utility in the discarded or seemingly obsolete. Think back to Mark Thomson's tinker networks —you know what I mean.

We found a quiet spot away from the crowds. "Let's start," I said. "How did you get involved with steampunk?"

"I always liked Jules Verne," Kimric replied. "In the '70s there was a Victorian movement, it was about costumes and mad adventures but for me Jules Verne was my constant. I was into the films but not just the Verne films, also *Dr. Doolittle* and the crazy ones like *Chitty Chitty Bang Bang*. I was fascinated by all the exposed machinery.

"But what really got me into it was this awesome club called Eli McFly's in San Jose, Cupertino. It was near Apple Computer. I did all this design work for it. It had a Victorian mad scientist theme.

"When was this?" I asked. You can take the historian out of the past, but never the inverse. I was, of course, busily sketching Bay Area patterns in the back of my head, working my way from the Merry Pranksters through to steampunk and to Burning Man. The whole thing just reeked of "not coincidence."

"This was in the '70s. Since then it was gutted out and turned into a jazz club in late '80s. Eli McFly's had an amazing stained glass window. It must have been like 25 feet long and the entire window was trimmed with t. rexes and pterodactyls, dinosaurs. When you came in the entrance you saw this Jurassic scene with this guy wearing a pith helmet, diving into the window. There was a speaker embedded in the glass. It looked like he was diving in the window and there were radiating ripples out from the speaker. That was just the entrance to the bar!" Kimric smiled.

"The bar was a time machine," he continued. "It was a glass cube topped by this bizarre blown neon. It looked like intestines. It glowed and flickered on and off. Floating around that were three bronze women with wings made out of shattered stained glass. Every night at midnight the time machine would turn on and you would see like shadows of the guy moving around inside. It looked like he was trying to return to our time and the bar was his lab. There was even a trilobite in a piece of sandstone with what appeared to be a badly corroded circuit board fossilized in with the trilobite, the detail."

Dipping into the Steampunk Academy's store of brass to fit his polarized lenses, Kimric steampunked Sunday's solar eclipse.

"Talk about playing to Cupertino..." I mused. This was in the heart of the personal computer revolution, which had in earlier turn been the heart of post-war counterculture. Eli McFly's patrons would have been a mix of 'heads and hardware jockeys. As I've said earlier in the chapter, there's little coincidence in this blend. Time machine lights attract a certain kind of moth.

"Oh my god, and the whole upper floor was all Victorian railing," Kimric remembered. "It went all the way around. It looked like an old German train station from the art deco period. Also hanging from the ceiling were these hammered metal zeppelins and the light fixtures were copper balls with rivets."

"How did you get involved with Burning Man?" I asked.

"I had designed this machine that could fire 2x4s at about 120 miles an hour," Kimric replied, as if that was no big deal. "It was like a baseball pitching machine with two semi-tractor trailer tires spinning, driven by a V8 engine. The whole thing held 25 2x4s. You could fire a 2x4 every 10 seconds. It would drive a 2x4 through a chain link fence. It was just crazy."

Kimric laughs as he continues, "We would put on shows in shipyards to show what the machine could do. At that time I was a part of the Cacaphony Society and was also doing stuff with Survival Research Labs. This was the late '80s. A friend of mine went to Burning Man in 1990 and told me I'd love it. Along with shooting 2x4s, I was also doing a lot of pyro stuff and he thought I'd fit right in. I knew Larry and when he found out, he wanted me to come a do some of my pyro stuff on the Burning Man. So I went."

The Larry who Kimric mentions is Larry Harvey, the founder of Burning Man. He'd been in and around the Haight during the Summer of Love (1967) and was both inspired and disillusioned by what he saw going on around him. By that time, the counterculture's infrastructure had been overloaded by an influx of young folks from all over. The dream was starting to crack under the strain. With Burning Man, Harvey created a dream for cultural makers. It's a dream of what could have happened. Burning Man is an economy based not on money and profit, but on art and heart—a place where near-infinite self-expression is the norm.

Kimric talked about the early days of Burning Man; he remembered a time when the "safeties were off." Creativity was the exclusive driving force. Guns and explosives? Sure! Just don't point 'em at anybody in anger. But those were the early days. Things have changed.

There's a kind of artistic, tribal balance out on the Playa, and it can be disrupted when personal expression crosses the line into imposition... or when somebody pulls a gun. Weapons are no longer permitted at the Burn (I'm personally grateful for this), and Burning Man now centers around a core set of ethics, expressed in the Ten Principles.

The Ten Principles of Burning Man

It's actually remarkable how well Burning Man's principles resonate with the ethics we've heard expressed in the steampunk community:

1. Radical Inclusion
2. Gifting
3. Decommodification
4. Radical Self-reliance
5. Radical self-Expression
6. Communal Effort
7. Civic Responsibility

8. Leaving No Trace
9. Participation
10. Immediacy

This is no coincidence either, but that's a longer tale, best held for another time.

"It's funny," Kimric continued. "Because I've been doing Burning Man so long I probably have as much experience burning down large structures as building them. It's an art. You have to do it in an engineered and aesthetic manner. You can't have one side burn, then the other—it falls in on itself. You have to engineer the burn so that it doesn't look intentional. It needs to look natural."

"You've got to have a pretty big crew," I remarked. Networks and community are starting to seem omnipresent.

"It's not really an organization," Kimric explained. "People feel affiliated with each other and there's a lot of insane amount of cross talk between all of us. It's a wide range of people. You'll find a guy who works at a crucible for a day job and he's also involved with this. But there's no center. The connections are like those flight maps you see of airplanes going between airports. It's a rat's nest of little strings."

"A bunch of relationships and interconnections." I said. It always comes down to the connections between people.

Time to pose the theory.

"I'm interested in the connections between Burning Man and the Pranksters and the other groups who were active in the '60s counter culture," I started. "I suspect there's a connection between all of them that also leads to steampunk."

"I think it's always been there," Kimric said. "That thing... that steampunk thing was always there. Back then it wasn't fashionable but now it's hip. That's really the only difference. It's great that it's bringing in money for people but interest will die off. In 10 years steampunk will be gone. What's sad to me is that it should be more DIY. People should learn to make more things than just costumes. I mean learn how to weld, learn how to solder, learn how to use a lathe!"

Yep, cool burns the edge off. But Kimric is one of the coals in our countercultural fire. He's going to go on burning no matter what happens; his eye is on the bigger picture, not on the quick flare-ups. He wants to *build* that fire up strong.

"When you weld something you feel it writhing around when you're welding it. As the bead pulls around it's a weird quasi-organic material and most people have never experienced that. I did shit like that when I was 8! I think people are worried about their kids getting hurt. They're worried they are going to burn their fingers. Well, I think it's going to happen either way. You either burn it when you're soldering when you are 8, or you burn it on a car engine when you're 25 because you don't know any better."

"There's an obsession with safety with children now that there wasn't before, there's good reason for it but..." I paused. As a parent my feelings on this topic are mixed. The world doesn't have bumpers and safety gates. On the other hand, I have a darn hard time even with the idea of my daughters anywhere near a truck-tire gun that shoots 2X4s through chain-link fences.

Kimric made an excellent point, though: "Yes and no, because it doesn't really create safety—it teaches people to have a complete unfamiliarity with their environment. There's no completely safe way to learn not to burn your finger. You don't want to learn by falling into a fire. You want to learn by burning the tip of your finger touching a soldering iron."

He's not wrong. Not by a longshot.

As we were wrapping up I told him that seeing the Neverwas Haul at the Burn in 2007 was what really started me on this whole adventure.

"Well," he responded, about to make my head explode (yeah, *again*), "if you want to come this year and help out, we can always use extra hands at camp. You basically live steampunk for a week. We'll have you pulling the head off an engine in the heat. It's hot, it's dirty and greasy but you'll be in the engine room of the Neverwas Haul."

There was no way on Earth I was going to turn down that invitation.

Sunday and Beyond: Neverwas a More Marvelous Thing

The last day at the faire was full of filming for the documentary and taking in as much of the scene as possible. While Byrd and his crew scrambled about in a fantastic frenzy of content-gathering, I took a different approach and sat comfortably in the shade of the Obtainium Works camp with Shannon O'Hare, the mad designer from whose inspired noggin sprung the wheeled wonder that is the Neverwas Haul. Doctor Professor Samuel Tweed, one of the Haul's able and industrious crew whose brilliant bartending keeps the Haul's other engine room—the parlor—well-lubricated, cracked open a little suitcase that turned into a portable

Byrd's-eye view of the Vintage Tomorrows crew shooting an interview with Shannon O'Hare. (photo courtesy of Byrd McDonald)

bar, drawing forth a frosty pair of gin & tonics. He presented them to us on a shining silver platter. Shannon's admiral's coat glittered with a mind-boggling accumulation of medals, ribbons, pins, and citations. Clearly this was a man of import, and yet I was addressed as the "guest of honor."

Shannon and I first met on the Playa in 2007 (Chapter 2), but I knew the Neverwas Haul had been there the year before. I was curious about how it all started. What I didn't expect in reply was an extra-dimensional answer to one of my central research questions.

"So, when did you build the Neverwas Haul?" I gracefully sipped at Dr. Tweed's delightfully frosty concoction. Ah, the wonders of quinine, juniper, and lime. Tonic indeed.

"We built the Neverwas in '06. My first Burning Man project was in '05."

"But that project in 2005 was something different? A clock tower, you said?"

"Right, that actually was another friend's project. He and Kimric were designing and building the clock tower. I came in as a master carpenter to build the tower. I brought in my own Victorian Gothic elements. There was no steampunk for us then. The word didn't exist in our vocabulary."

"I hear that a lot, actually," I added.

"What's fascinating is that we went out to Burning Man, I wore this uniform and this hat with this badge." He pushes forward one of nigh-unto-a-hundred badges that adorn the front of his majestic admiral's coat. "This was the regularity badge and we were the time regulators."

A solid start for a steampunk story.

"The idea was that we had created this clock on the Playa and somehow had created a chromatic vortex anomaly that threatened to open a portal to all other dimensions. We had to destroy the clock exactly at midnight, to prevent the portal from opening."

"And if you failed, and the portal opened?" I asked.

"Reality would be forever torn asunder," he smiled. "We were actually delayed so we didn't burn the clock on time."

"Oh no," I said.

"So as a result, the Neverwas came through that anomaly..."

"Wait! 2005, 2006... that's when I've traced the start of the big surge in steampunk culture!" I near jumped out of my chair.

Shannon cocked his eye in a knowing trickster's grin. "That's right."

"Because you didn't burn the clock..." I trailed off.

He sipped his drink. "In time, yes. We didn't burn the clock in time."

Whoa.... A technological tipping point with powerful historical parallels, a countercultural nexus, *and* an interdimensional rift? What more evidence do I need that something big happened to steampunk—and our future—in the mid 2000s?

"So, the clock tower was truly the first steampunk large-format piece," he continued. "Before that there was no real steampunk on the Playa. The Kinetic Steam Works fellows started the following year—the same time we put in for the Neverwas. All of a sudden everybody went, 'Oh we can do Victorian? I didn't know we could do Victorian.' Before that it was mostly just the totem Burning Man style, which was kind of like trash art, steel, but there really was no Victoriana, there was none of that style."

Shannon continued, "We were sitting around camp after the clocktower burn, talking about what we were going to do next. I had just finished reading *Snow Crash*. Neal Stephenson was writing about the virtual world where you are not limited by laws of physics. You could write a program in which you had a three-story Victorian house on tank treads and you could drive it at 80 miles per hour. I thought, well, Burning Man is like that virtual world. We aren't limited by the laws of physics here. So I said, *Let's build a three-story Victorian house on tank treads.* And we did. That's how we decided to build the Neverwas Haul."

The real source of 21st Century steampunk? The Clocktower on site at Burning Man 2005. (photo courtesy of Kimric Smythe)

"I was worried that everyone at Burning Man was going to resist the Neverwas," Shannon said.

"Why?" I asked.

"The Neverwas came through that anomaly..." Shannon and the Haul. Anomaly not pictured. (photo courtesy of Andy Pischalnikoff)

"Because we were bringing this Victorian design to the savages at Burning Man," he explained. "I figured they'd rebel against it."

"And what happened?" I said, partly knowing the answer already.

"We threw a tea party," he smiled. "And everyone showed up. Everybody loved the soiree and the whole big event."

Then it hit me. Massive iconic vehicles defying the laws of physics? Mobile magic on a grand scale? A spontaneous tea party? This is *The Bus*! Of course! The Neverwas Haul is steampunk's Furthur!

FURTHUR INTO THE NEVERWAS!

Here's a bit from *The Electric Kool-Aid Acid Test* to give you a sense of what I'm talking about:

"They took a test run up into northern California and right away this wild-looking thing with wild-looking people was great for stirring up consternation and vague befuddling resentment among the citizens. The Pranksters were out among them now and it was exhilarating—look at the mothers staring!—and there was going to be holy terror in the land. But there would also be people who would look up out of their poor work-a-daddy lives in some town, some old guy, somebody's stenographer, and see this bus and register... delight, or just pure open-invitation wonder. Either way, the Intrepid Travelers figured, there was hope for these people. They weren't totally turned off. The bus had great possibilities for altering the usual order of things."

— TOM WOLFE

The Electric Kool-Aid Acid Test (1968)

For any who've seen it rumble past, or have clambored around its iconoclasticly anachronistic interior to the chap-hop backdrop pumping out of the horn speaker of Kevin O'Hare's (there are three brothers O'Hare, all involved with the Haul) Edison MP-3 player, the Neverwas Haul plays the same role. It brings its own adventure with it wherever it goes. Just like the Pranksters drew everything around them into their "movie," the Haul and its crew sweep everything around it into a past that, well, never was. At Burning Man this year, we drove out into the middle of the open Playa and served proper tea (milk and lumps of sugar, china cups and saucers, the works) to folks who happened by. We also staged an invasion of the "Steampunk Saloon" and conducted a children's art tour. The Haul *is* adventure.

It rapidly became clear to me, as I was assimilated into the crew, that like the Pranksters and their bus, the Neverwas Haul operates in symbiosis with its crew. This runs even deeper than the connection between Neal Cassady and the bus's engine, though, because it literally does take a family to drive a house. There's no rear-view mirrors, no airbags, no safety belts. The Haul steers like a ship and has the visibility and turning radius of, well, a Victorian house. It requires at least a half-dozen people to drive the thing safely—one at the helm, another on a "deadman switch" which instantly stops the Haul when released, and multiple spotters and wing-folks, clearing the way and extending the pilot's visibility.

Driving the Neverwas Haul is an experience like no other—requiring an extension of the senses to incorporate the entire visible and audible environment, as well as (I can think of no better way to say it) developing a feel for the Haul itself by becoming, in a way, a part of the machine yourself. Months after my brief stint at the helm of the Neverwas Haul (as I sit in my local coffee shop, preparing this final draft for handoff to our copy editor) I can still recall the expansive feeling in my chest, the way my legs felt an extension of the gently rocking deck, the heightening of my senses as I drew information not only from my own eyes and ears but from the crew around me. It's a symphonic experience, so unlike the solitary experience of driving a car as to be completely incomparable. When pressed for a brief answer, the best I can muster is: "It's like driving a house. You, er... You know what I mean, right?"

I still reel in bliss when I think back to this moment. I doubt that will ever change. I got to drive the Neverwas Haul. Burning Man 2012. (photo courtesy of Samuel Coniglio)

I could, and no doubt will, write about my experience aboard the Neverwas Haul at much greater length, but you've much more past and future to explore and I will detain you no longer on this point. In sum, the Neverwas Haul is an artifact of the imagination, an icon like Kesey's bus that transports the world around it into another frame of mind; inextricable from the human beings that give it life, it is not only a living embodiment of steampunk—it is now a member of my family, just as they are.

Bully for expecting the expected (and not forgetting the rest). My trip to Maker Faire really did lead me to something truly spectacular. I'd come full circle.

I stepped out the door of my Hobbit hole long ago, journeying into the past and out to the cultural fringe. My adventure took me from the American Civil War to beatniks and hippies, from Burning Man to steampunk to the septic tink-tank, Maker Faire, and back to the wonderment that is the Neverwas Haul. I'm a steampunk now—ready for anything—so, unlike the otherwise estimable Bilbo Baggins, I never forget my pocket handkerchief.

Through all this I had not only found my steampunk Snark, but bred it. No longer a Snark hunt, my journey became a travelling Snark emporium, with imagination at the wheel and a world of possibility in my head *and in my hands.*

Possibility (photo courtesy of Andy Pischalnikoff)

Pop Goes Steampunk

The steampunk world is rocked with the news that teen idol and global pop phenomenon Justin Bieber released a holiday music video that looks remarkably like steampunk. Is steampunk making its way into the mainstream? What is the relationship between music and technology? How is Justin Bieber like the Beatles? Culture change is messy!

As you know from Chapter 1, this entire project started when James and I were having a beer and he told me that he thought that steampunk was gaining in so much popularity that it was an indicator of something larger going on in our popular culture. As he explained, subculture like the Beats of the '50s and the hippies of the '60s were indicators that something much larger was going on beneath our entire culture.

That's how it all started, and during the holiday season of 2011 we discovered that our cultural historian could not have been more right.

Steampunk Is Coming to Town

Brian David Johnson (Los Angeles, California)

On Wednesday December 7th, 2011, Rob Beschizza posted a video to the wildly popular blog Boing Boing. The post was titled "Justin Bieber's Steampunk Christmas Video." Directly under the video the only comment by Rob was, "Cory this is your fault," a reference to Cory Doctorow's relentless reporting and support of steampunk.

I talked with Cory about this and he said that when Bieber's video came out he really wasn't paying attention. He was on a book tour in Germany at the time.

"Makes sense," he said. "It's an aesthetic now... a look."

He paused then said, "I guess I do talk about steampunk a lot."

The time has come: let's talk about Justin Bieber and "Santa Claus is coming to Town."

For those of you who are not under the age of twenty or who don't have children who are, let me explain. In 2011, Justin Bieber pretty much owned the world. He had actually been there since March of 2010 when he released *Baby*, an infectious pop delight. The song was a worldwide hit and the music video ranked as one of the most watched videos on YouTube. The fresh-faced Canadian could do everything: he could sing, he could dance, and he seemed like a nice kid. Boom! Global pop star. Fifteen million in album sales to date, leading many people to start talking about Bieber Nation, thus confirming it was Justin Bieber's world and the rest of us are just renting out space.

The song "Santa Claus is Coming to Town" was from his holiday album, *Under the Mistletoe*, also released in 2011. The song was featured on the soundtrack for the animated holiday movie *Arthur, Christmas*, a wholesome story about Santa's son, who must complete a mission before Christmas morning. The film was a moderate box office success making $150 million worldwide.

Bieber's music video, dubbed the "Arthur Christmas Version" has all the steampunk trappings; gears, steam, goggles, bowler hats, automatons, and Bieber even has a mechanical arm. Don't get me wrong, it's not all clockwork magnificence. There's also a lot of break dancing, drum playing (on a distinctly steampunk drum set), and a cavalcade of synchronized dancers. 'Twas good pop fun.

Was this video the evidence that steampunk had made its way up into the mainstream? It's not that Justin Bieber wants to be a steampunk; I'm sure he's quite happy being a pop star. Was it something about the aesthetic, the look of it that appealed to the artist and his team?

Justin Bieber in "Santa Claus is Coming to Town" (photo courtesy of Charles Oliver/Alice Brooks)

Let's ask them!

I called Scott Braun. He's known to everyone as Scooter. Technically Scooter is Justin Bieber's manager but it's not as official as all that. As Scooter explained to me it's a small team, more like a family. There is no large entourage of people fastidiously circling around the pop icon. They keep it pretty small and definitely low key. Many people have likened the relationship between Scooter and Bieber to that of legendary power team Elvis Presley and Colonel Tom Parker. Not too shabby company to keep.

I called Scooter midweek when we were both dashing between meetings and appointments. I had expected to have a quick chat to set up a time for another call so that we could talk about maybe doing an interview that possibly might get into the subject of steampunk and Bieber's video.

"I'm stoked," Scooter yelled into his cell phone as he drove into the Hollywood Hills. "Let's do it right now!"

Really? Holy cow! Scooter's enthusiasm and laid back approach was as refreshing as it was unexpected. You have to remember he manages this cute Canadian kid who owns the world. Not only that he also manages an impressive slate of up-and-comers like The Wanted, Asher Roth, Mike Posner, Cody Simpson, and Carly Rae Jepsen. And he wants to geek out with me on steampunk and its implications on technology and pop culture. Really? The world is never what you expect it to be. Let's do it!

To start off, I need to tell you that Scooter is really kind of a futurist, too. He recognizes the indicators of culture change and their implications on technology.

In the entertainment industry he is sometimes called "the future of music guy." Early on in his meteoric career, Scooter understood the strong relationship between music and technology. (There's a great story about how he just missed out on investing in Facebook before anyone had really heard of Mark Zuckerberg and "friending.")

"If you want to keep an eye out for real change in popular culture you have to look at music," he told me. "Music is where it always shows up first."

It makes sense that trends would come and go quickly through the music industry. It's a relentless culture machine that keeps changing and exploring the "new." Pop songs and music videos and even fashions have an incredibly short shelf life. If you are over thirty all you need to do is go talk to a teenager and you will quickly learn that you have no idea what is really going on in the world. It's humbling and a lot of fun. If you don't have one, go out and get yourself a 13 year old mentor. You'd be surprised how much you'll learn.

Scooter went on to explain that he saw a strong bond between music and technology. He saw them both as culture; technology culture and music culture feeding into one another. If there was any new technology or technology trend, music would pick it up and play with it for a little bit to see if it was cool. If it wasn't cool it would be dropped. If it was cool and popular and resonated with people then it would "break out" or start to "trend" (these are my words, not Scooter's—he's much too relaxed for jargon). If the tech catches on then it starts to show up in more videos; it shows up on web sites and permeates social networks. It's not exact. It's messy and fun and the inexact science of culture change.

The day before I finished this chapter I was taking a flight up the west coast of the US. Before jumping on the plane, I swung past the newsstand and what did I see... Bieber! Decked out in a suit on the cover of the June 4th, 2012 edition of Forbes magazine was Justin Bieber with the headline "The Most Surprising Business Story You'll Read This Year." The lengthy article explained the partnership between Bieber and Scooter, actually calling Scooter "Bieber's Business Brain." Turns out over the last few years Bieber and Scooter have been investing heavily and successfully in technology companies like Spotify, Tinychat and Airtime. So when Scooter talks about the relationship between music and technology, he knows what he's talking about.

"So, why steampunk?" I asked. "How did you and your team come up with the idea of doing the *Santa Claus is Coming to Town* video in a steampunk theme?"

"It was cool," Scooter replied. "At my 30th birthday party I had seen a dance troupe called Lucent Dossier and they had a steampunk show and I thought it was cool. So I looked into it and though it was perfect for the video. You have to remember that the song was also used in the movie Arthur Christmas which has a theme of mixing the old and the new. Have you seen it?"

I hadn't but I have now. I was in London when *Arthur Christmas* opened and Regents Street was decked out with the inflatable character of Arthur and massive clusters of immense, glittering snowflakes. It was very pretty. I was trying to get a haircut. I was far too serious for all that holiday fun. My loss.

"The theme of *Arthur Christmas* is about the old and the new," Scooter continued. "So I thought it would be perfect to use steampunk for the look of the video. It's the old and the new together."

Ok, so here all of my preconceived notions of pop frivolity have been dashed on the rocks of serious thought. I pushed on. So what did it mean? Why was the video so popular? Why are people so drawn to the aesthetic... to the look of it?

"I think steampunk is a perfect mix between the old and the new," Scooter explained. "People are looking for some way to process all this new technology. Steampunk gives people a way to understand new technology. When you put it in the trappings of the past people feel more comfortable with it. It's easier to accept. I think people are searching for that. They want new technology but they also want to be comfortable with it. Steampunk does that for people."

An Aside from the Historian

Pop is, well, pop. It cannot be all things, nor can it always be expected to retain a critical edge. As China might say, that's not really its job. That being said, I would be remiss in pointing out that music video evidence to the contrary, children in turn-of-the-century factories did not bust funky dance moves on the job. This is not because hip-hop wasn't invented yet (this is steampunk, for crying out loud), but rather because they'd have fallen into the machines, lost a limb, and been fired because the job required two hands and two legs.

I'm not suggesting that Justin and Scooter ought to have removed kids from the video. It was Santa's factory floor after all. I just feel obliged to offer a "make it better" example in the ministrations of our favorite historical robot:

Textile mill workers with Boilerplate, early 1900s (photo courtesy of Paul Guinan and Lewis Hine)

Scooter saw steampunk as a way for people to process all these new technologies that he loved. Music culture was the onramp for the technology that was coming. It made the technology cool. Music and popular culture were the trappings around that technology. It made it cool or at least presented in a way for people to see if it might be cool.

And steampunk had won! Maybe "winning" isn't quite the right way to frame cultural change, but this was *big*. Steampunk stepped up on a much bigger stage that day than it had ever taken before (in terms of broad cultural reach, Justin's audience makes an article in the Fashion section of the *New York Times* pale in comparison).

(photo courtesy of Charles Oliver/Alice Brooks)

STEAMPUNK GREW THREE SIZES THAT DAY

> *And what happened, then? Well, in Whoville they say - that the Grinch's small heart grew three sizes that day. And then - the true meaning of Christmas came through, and the Grinch found the strength of* **ten** *Grinches, plus two!*
>
> — **DR. SEUSS**
>
> *How the Grinch Stole Christmas* (1966)

However you frame it, though, one thing's for certain: steampunk gave Scooter and Justin a visual language that allowed them to express something new and meaningful. To spin another Christmas tune, the conversation about our future and our past grew like the Grinch's heart that day. Steampunk may just have found the strength of ten steampunks plus two!

"Do you remember the drum set from the video?" Scooter asked.

I did, very much. I actually mentioned it above. It's an amazing thing.

"They were going to get rid of it after the video shoot," Scooter said with indignation. "I couldn't let them take it. It was too cool. I put it in my house. It's the most popular thing in my house. When people come over that's all they can talk about. I kept the glove too."

Scooter picked up on steampunk for two reasons. First, he thought it was cool. Second, it was a bridge between the past and the future that made people feel comfortable. This supported the idea that steampunk is having a broadening influence on our culture and how we are thinking about our relationship with technology. All the trappings, the goggles, the hats, the gears are the tip of the iceberg. Literally the reason why the most famous singer in the world picked steampunk as the look of his video was because it was a bridge between the old and the new.

This has happened before. When James and I were talking about Bieber, our cultural historian pointed out that this is not the first time this has happened. The Beatles did the exact same thing with the Ken Kesey and the Merry Pranksters. You remember the Merry Pranksters from Chapter 2.

In 1967, the Beatles owned the world. The British band was loved by critics and adored by fans. That year they released *Sgt. Pepper's Lonely Hearts Club Band*, a psychedelic rock album that was a massive success. The Beatles followed up their hit with a film and album titled *The Magical Mystery Tour*. The costumes and personas in the two albums and the film are taken from Ken Kesey and his Merry Pranksters who were making all kinds of trouble in the America at the time, exploring their own subculture. The style of the Beatles costumes borrow from the style of the Pranksters. The super group even went so far as to take on the style of the Pranksters' infamous bus Further with their Magical Mystery Tour Bus. The Beatles borrowed the look and ideas of an emerging subculture and put it in their videos.

It's a different world today, but Justin Bieber sure has a big chunk of it. It's not a huge stretch to draw a parallel: Justin Bieber is to steampunk as the Beatles are to the Merry Pranksters of the '60s.

But maybe all of this use of steampunk might just be a fluke. Maybe Scooter was a secret steampunk and he just didn't want to tell me. Maybe even Bieber himself carried around a pocket watch, hiding it from his throngs of adoring fans. Perhaps in 2011 we just saw a random appropriation of culture hauled into the mainstream for fun. Not so fast.

On February 1st 2012, the Internet site Buzz Feed posted a new music video "Niki Minaj's New Steampunk Video" with the tagline "The Biebs had a steampunk Christmas, and now it's Nicki's turn. " Oh goodness.

Justin Bieber is a wholesome teen-idol and global phenomenon but Nicki Minaj is something else all together. Minaj is an edgy performer that pushes the boundaries with alter egos and influences from Marilyn Manson to Lil Wayne.

The video is a collaboration with David Guetta titled "Turn Me On." Whereas "Santa Claus is Coming to Town" had all the trappings of steampunk, "Turn Me On" throws itself decadently and magnificently into the genre. Minaj is an automaton built by a goggled inventor, which moves through a sexualized and threatening world of other automatons that pursue her. To date the video has nearly 76 million views on YouTube.

As we head into publication, steampunk continues its roll into popular culture. Tyra Banks orchestrated a steampunk photo shoot for an episode of *America's Next Top Model*, Prada's Fall/Winter menswear collection featured the likes of Gary Oldman and Willem Dafoe in high steampunk style, and Tori Spelling planned a network TV steampunk wedding for a Los Angeles couple. We'll spare you the rest of the details here, but it's darn clear: steampunk has gone pop.

Culture change here we come...

Not even a rainy day in L.A. can stop me from bringing our insights to bear on questions of the future. (photo courtesy of Byrd McDonald)

A Note from the Futurist

Brian David Johnson (Portland, OR)

We have journeyed through the depths, intricacies and tangents of steampunk culture and now we find ourselves standing on the other side. Oh what a trip it's been! But what does it all mean? What can steampunk teach us about the future of technology?

One of the most important things I learned from James and our research was that this is real. The view through Byrd's camera also confirmed that steampunk is an indication of culture change. It is telling us something about our broader popular culture. More importantly for me as a futurist, steampunk is telling us something about the future of technology. It's evidence that people's relationships with technology are changing. People want something different, they expect more from the devices that have become such a part of our daily lives.

Steampunk showed us that people want want their technology to have a sense of humor, a sense of history, and finally a sense of humanity. People want more humanity both in *and* around *their devices.*

This is the final and most important thing we learned: in the end it's really all about people. So we talked to a few more! We talked with a paragon in the world of comedy, a leading anthropologist working in the technology industry, a world renowned designer, and even another futurist. We ask them what they think about technology, and how it can have more humor, history, and humanity (AKA the "big three").

This last act of the Vintage Tomorrows story examines the meaning of what we learned. We explore the science behind the big three and look to what we can do with what we learned. How do we apply it? What does it mean for the future of technology?

In the end we're the ones who build the future, and (like the past!) we build it out of the stuff of imagination.

13

Humor Is the New Killer App

People are funny. It only makes sense that we would want our technology to have a sense of humor. No one likes a person that can't take a joke. Why would we tolerate a computer that can't take a little ribbing now and again? We want our technology to get us, to understand who we are as people. What better way to express this than to get our jokes; to think we are funny and to make us laugh? Brian and Byrd travel to LA to learn more about humor and how we might be able to apply it to the future of technology.

Siri, Where's the Closest Comedy Club?

Brian David Johnson (Los Angeles, California)

> *The popularity of Siri shows that a digital assistant needs more than just intelligence to succeed; it also needs tact, charm, and surprisingly, wit. Siri gets the joke and plays along. Thus it has a clever answer for just about any curveball thrown at it. It even varies its responses, a trick that makes it seem eerily human at time.*
>
> **— WILL KNIGHT**
>
> MIT's *Technology Review* (June 2012)

LATE IN 2011 a perfect example of humor being used in technology dropped right into our hands... literally. Apple released its iPhone 4S with the voice-activated personal assistant, Siri, and we were floored. Increasingly as the reviews rolled in about Siri most of the reviews didn't talk about the application's ability to find you a great place to get sushi. Even MIT *Technology Review* online editor Will Knight recognized the magic of Apple's Siri in his article about social intelligence. What most people talked about were the jokes. Siri was funny. Siri had a sense of humor!

But Siri is just an early example. It wasn't that Siri had a sense of humor; the programmers over at Apple had a sense of humor and put it into their assistant. The jokes and the witty little comebacks are all canned. We aren't putting down Siri. We're just saying it's still in its early days.

So if we were going to give our technology a sense of humor then how would we do that? And why do people want their technology to be funny? Humor is pretty complicated. What kind of humor do they want? Jokes? Wit? Satire? Irony? Snark? The list goes on and on. What do they want it to *do*?

To find out Brian, Byrd, and the crew went to one of the centers of prediction of comedy in the world: Los Angeles (and yes we are holding back all the jokes that could come from this statement... I mean come on... that's a softball!). And then we did a little digging into the theory and history of the relationship between humor and technology in the past. Maybe we didn't get all the answers, but it's fascinating what humor reveals. Turns out that "funny" is a really complicated thing.

What's This Humor Business All About? Jaime Masada

> *"The greatest thing you could do for mankind is to bring smile to a person's face."*
>
> **— JAMIE'S DAD**

When Jamie Masada saw his first comedy at age six be laughed so hard he peed in his pants. How's that for an opening sentence? I figure if we're going to do a chapter on humor, we might as well start off with a bang.

It's weird when it's rainy and overcast in L.A. James, Byrd, and I are from the Pacific Northwest so we're pretty comfortable with rain. But rain in L.A. just seems *wrong*. I've been coming to L.A. for over 20 years and the weather is almost always nice. You have to give that to Angelinos—the traffic might be ridiculous and the smog might be troubling, but the weather, well, you can't beat it.

Byrd and I had come to L.A. to chat with Jamie Masada—the founder of the world famous comedy club The Laugh Factory. Jamie has made comedy his life but he's also made the well-being of people all over the planet his mission. Jamie's been in the comedy business for decades—producing, consulting, writing, publishing, performing, and doing everything he can to make the world laugh. Laughter is Jamie's business. He's given countless comics their break in the business but usually all he wants to talk about is how he can get you to help him out with his next charity event. I think he might be an angel.

The drive from the airport to Sunset Boulevard was gray and damp. Coming down Highland Ave and passing the Hollywood Bowl, I really missed that sunny L.A. feeling. Byrd was filming as we drove, asking me about Jamie and my history with him. This was my first chat with Jamie but many more followed. The more we talked the more I discovered the depths of his insights into humor and the human race. To be truthful, he's kind of a revolutionary. He thinks that comedy and technology together can save the world. Technology allows people access to more people all over the world and if we use that technology to tell jokes, to get people to laugh then we won't kill each other anymore. Not a bad vision for the future.

Jamie's office is just off of Sunset Boulevard in west Los Angeles. We parked in the small lot behind the house and went up to the back door. And it is just that, it's a back door of a house, but what a house. Jamie's office is in Groucho Marx's old house. His comedy club is in Marx's old private theater. How's *that* for history?

Jamie Masada and Brian David Johnson stand on Groucho Marx's rug (photo courtesy of Byrd McDonald)

"You're standing on Groucho Marx's rug!" Jamie called to me as he came down the steps. I looked down and felt both in awe and a little worried that I had done something wrong. I started to move but Jamie smiled and gave me a hug. "It's OK, my friend Groucho would have wanted you to stand on his rug."

Yes, Jamie actually talks like this and he's completely sincere. His accent is thick and a little hard to place. It's a nutty mix of Farsi, Hebrew, and show biz banter. He sounds like no one I've ever met.

It's safe to say that Jamie Masada knows comedy. His club is located at 8001 on the Sunset Strip and has been a landmark and beacon for comics for decades. On the day we were there they were holding auditions for their legendary amateur's night; the line stretched around the corner. When the would-be comedians saw Jamie they tried not to stare but a few couldn't keep their cool. They descended on Jamie, but he greeted them with his usual good-natured enthusiasm.

"You gotta kill them tonight my man, you gotta kill," he smiled and led us into the front door.

Embarrassingly this was my first time in the club. I think Byrd had been there before. Like many things in life, when you actually get there it's always surprising how small really famous places are. But I was quickly distracted by the memorabilia. Holy cow! Tim Allen. Red Buttons. Bob Hope and on and on, all safely shut in small dark wood cases with dainty accent lights. Jamie is a collector of the history of American comedy and all his clubs are little mini museums.

The roster at the Laugh Factory is a who's who of decades of comedic genius (and this is the SHORTENED list!):

Adam Sandler, Dane Cook, Tim Allen, Jim Carrey, Jerry Seinfeld, Ellen DeGeneres, Chris Rock, David Spade, Richard Pryor, Rodney Dangerfield, Eddie Murphy, Jon Lovitz, Robin Williams, Paul Rodriguez, Damon Wayans, Ray Romano, Bob Saget, Chris Tucker, Brad Garrett, Dave Chapelle, Martin Lawrence, Carlos Mencia, David Alan Grier, David Letterman, Kathy Griffin, Harland Williams, Kevin Nealon, Tom Arnold, Steve Martin, Jimmy Fallon, Freddy Soto, Roseanne Barr, Will Ferrell, Jamie Foxx, Paul Mooney, Ricky Gervais, Michael Richards, and George Carlin.

Jamie took us upstairs saying, "Later on we need to get you on stage my friend. This is your lucky day my friend!" I heard Byrd chuckle behind the camera as he followed us.

The back offices of the Laugh Factory are both historic and mazelike, revealing the age of the building. It feels like part comedy temple and part hectic show office. We sat down in Jamie's office overlooking Sunset Boulevard, the crew got us miked up, and we started chatting.

With the smiling faces of Laurel and Hardy and a few ceramic clowns looking down on us, Jamie told me two great stories about humor and comedy. One is about the first time his father introduced him to comedy, the second is the first time he went on stage to do stand-up in the United States.

First we start with the peeing in the pants story.

When Jamie was about six or seven years old his father told him that if he was a good boy and minded his parents that he would take the young Jamie to see a *moving picture*. Jamie grew up poor in Iran and had never seen a moving picture before. His family didn't have a TV so Jamie was excited. He was a good boy and then the day came that his father said he would take him to see the moving picture.

Father and son walked a long way out of their village to a larger town nearby. His father didn't have the money to take Jamie to the movies but he did take him to a TV repair shop. In the window was this little black and white TV. The shop owner had left it on and it was showing a Three Stooges movie.

Jamie couldn't believe his eyes; they popped out of his head. He had never seen the little people in the little box before. He wondered to himself: *How did they get those little people in the little box to do all of that?*

Then his father started doing all the voices and the sound effects for the Stooges. Because the TV was inside the shop the pair couldn't hear the sound. He told Jamie what was happening and really hammed up the sound effects. *Bang! Bang! Yoinks! Pow!*

Little Jamie started laughing. By this time they were sitting on the steps in front of the store and Jamie couldn't stop laughing. He had never laughed so hard before, so he peed in his pants. That trip and the stories his father told Jamie about the Three Stooges were the greatest gifts the father had ever gave him. They both laughed for a long time together; neither ever forgot that trip.

After that first trip to the TV repair shop, Jamie and his father would return once a month at night. After the store had closed, they would sit across the street and watch more Three Stooges movies, along with other classics like Laurel and Hardy. Jamie and his father would watch the movies and do the sound effects and dialogue together. That's how Jamie Masada was introduced to the power of comedy and laughter.

When Jamie first came to the United States he was very poor and looking for any job. He knew he wanted to be around comedians and comedy clubs so he was prepared to do anything at all. He'd be a dishwasher, glass washer, ball boy—he didn't care. (I didn't ask what the ball boy would actually do at a comedy club, but Jamie was on a roll so I couldn't stop him—Jamie's like that when he's on a roll). You get the point—he'd do anything.

So Jamie went into this one club and one of their comedians didn't show up. The club owner asked Jamie, "Can you tell jokes?"

"Yeah, I've told jokes before," he replied. "I've told jokes at weddings and different family functions with my father. My father taught me a lot of jokes."

"Great," the club owner said. "You go on in five minutes."

Jamie was petrified. He was only fifteen years old at the time but he was determined to work in comedy. The comedian before him finished up and Jamie was announced. As he walked to the stage he could see that there were about thirty to forty people in the audience but Jamie didn't let his nerves get to him. Right when he got to the stage he started telling jokes, one after the other. He told jokes that his father had taught him. He told jokes that he and his father had told together before. He got into it. He started having a good time.

From the stage Jamie couldn't see anyone. The lights were too bright and he was too excited. Now the thing that Jamie forgot during his big break in a comedy club was that he was in America. You see, he had never told a joke in English before. The only jokes he knew were in Farsi and Hebrew.

The greatest part about the story was that audience started laughing. They were laughing along with Jamie. They thought it was a joke and Jamie was having such a good time that they laughed along with him.

"Laughter is a universal language," Jamie told me. "It's like music. You don't have to understand it. Everyone knows funny. If somebody starts laughing, then somebody else will start laughing, then somebody else... it's catching."

After his stories I explained the work that James and I had been doing and what we had learned. I wanted to talk to him about this humor idea.

"So Jamie," I started. "This might sound like an odd question, but what do you think it means to give technology a sense of humor?"

"I don't think that's an odd question at all," Jamie started.

"But how do we do it?" I continued. "Why would people want their technology to have a sense of humor?"

"That's an easy one. Laughter is what brings people together," Jamie started. "Here's a true story I saw the other day. I went to a restaurant with three of my friends. I saw this kid sitting off to the side all by himself. He was eating alone but laughing really, really hard. He was watching his smartphone and laughing so hard that my curiosity got the better of me. I had to know what he was watching.

"I went up to him and said, 'I'm sorry to interrupt you. But what are you watching?'"

"The kid looked back at me petrified." Jamie mugged a frightened face. "Turns out this poor guy is from Korea and doesn't speak any English. But he showed me what he was watching. You want to know what it was? He was watching Jim Carrey's physical comedy and he was laughing really hard. It didn't matter that he didn't understand what Jim was saying. But he was watching a smartphone, the smartphone was keeping him company, making him laugh. This poor kid is all by himself from Korea and it's his phone that's his friend.

"I thought to myself, *Oh my God this is what brings people together*. It's not just the humor and the comedy but it's what you just said about technology. The way people relate to technology is through humor. As soon as you can get comedy on you device, your smartphone or your iPad or your laptop, if you can get something humorous and you laugh, then that technology becomes your friend." Jamie

Director Byrd McDonald shooting the Vintage Tomorrows documentary (photo courtesy of Alan Winston)

held up his smartphone. "You can't just consider it a phone or a computer anymore; you think of it like a member of your family, it's your friend. Maybe it gives you a little joke or a cartoon, doesn't matter, the important thing is that it made you laugh and you forget about the technology."

"What do you think it means that people want this kind of relationship with their technology?" I asked.

Jamie thought for a moment then replied, "People want the technology to relate to them and they want to relate to the technology. It's like that poor guy from Korea in the restaurant, he's sitting down all by himself and eating and laughing. He could have dinner with his family or his friends or a date but he's not. He's having dinner with his smartphone and Jim Carrey and he's laughing his head off. The phone is his friend. Technology has slowly, slowly become like another human being. People have a relationship with their technology now. They can understand it. And if that technology has a sense of humor, well then that's the best thing it can have."

The interview finished up. Jamie and I continued to chat about when we might come down to the club and see a show. He said all of us had to come that night, but we couldn't. We needed to fly back up the coast.

"Ok, ok, my friend," Jamie looked at me, Byrd, and the crew sideways. "I get you next time."

After leading us out, Jamie hugged Byrd then me and with a smile said he'd see us soon.

"He's amazing!" Byrd said from behind the camera as he filmed and I drove us back to the airport.

"Yeah," I replied, "he's pretty amazing. I had no idea he was such a revolutionary."

"I think he wants to save the world," Byrd said and panned the camera out the car window.

> *A person without a sense of humor is like a wagon without springs, jolted by every pebble in the road.*
>
> — **HENRY WARD BEECHER**

Humor Is Like Sex (Err, at Least So Say the Experts)

> *Humor can be dissected as a frog can, but the thing dies in the process and the innards are discouraging to any but the pure scientific mind.*
>
> — **E.B. WHITE**
>
> reprinted in *Humor 101* (2000)

When you start doing research about humor you quickly learn a few things. First, there's really not that much work that's been done on the subject. Second, you learn that among the work that has been done the experts agree that they will never agree. It seems that humor is a pretty elusive thing.

Mitch Earleywin's book, aptly titled *Humor 101*, was a good place to start. Early on he admits, "Psychologists and other cynics often suggest that humor can't be studied at all. They lump a funny thing in with strange bedfellows like love and pornography. They know what's funny when they see it, but it can't be pinned down."

Mitch isn't the only one. John Kachuba's collection *How to Write Funny* gathers together advice and wisdom from funny people like Dave Berry, P.J. O'Roark, Jennifer Cruise, Tom Bodet, Bill Bryson, Joe R. Landsdale, and Roy Blunt Jr. Most of them say that comedy is hard but comedy is a way of life. In his introduction John observes, "Humor is like sex: Everybody knows exactly what it is, but no two people ever agree. College professors have written dozens of books offering philosophical, psychological, sociological, historic, poststructuralist and even biological theories of humor. They all come to the same conclusion: We don't quite understand humor yet, and we require generous funding for further research."

Over and over it seems that humor is elusive and the moment anyone comes close to pinning it down, it slips away, changes and becomes something different. This is problematic from an engineering standpoint. If we are going to design more humor into our technology and devices we need to have a set of requirements to design to. Essentially we need a target. We need a way to understand what humor is so that we can use it to design the hardware, software, and systems that are behind our technology.

Jim Holt gives us a clue to where we could look in his 2008 book *Stop Me If You've Heard This: A History and Philosophy of Jokes*. He noticed that "While there can be laughter without humor—tickling, embarrassment, nitrous oxide, and vengeful exultation have been known to bring it forth—there cannot be humor without laughter. That, at any rate, is what contemporary philosophers think. *The propensity of the state of amusement to issue in laughter is arguably what is essential to its identity*, we read under "Humour" in the Routledge Encyclopedia of Philosophy. But laughter is physical. You need to have a body to do it. So, if the philosophers are right, purely spiritual beings couldn't really *get* a joke."

Jim makes a great point. Humor is not some esoteric thing. Humor is about people; it's about having a body. You have to *think* something or someone is funny and then you need to laugh, or chuckle (or in the case of Jamie, pee your pants). It seems there is something uniquely physical about humor, and to understand this maybe we should stop looking just at humans. What if humans are not the only ones who can have a sense of humor (HINT: this is a good thing if we are trying to give our devices a sense of humor). If we can examine humor as something that is both definable and yet not uniquely human it could give us a way to come up with ideas for how we get that sense of humor into our technology. Maybe we should look to... RATS!?

Jaak Panksepp is an Estonian-born American psychologist, psycho-biologist, and neuroscientist. He is currently based at the Baily Endowed Chair of Animal Well-Being Science for the Department of Veterinary and Comparative Anatomy, Pharmacology, and Physiology at Washington State University's College of Veterinary Medicine—but he's most widely known as the man who makes rats laugh.

In 1996, working with two other researchers, Brian Knutson and Jeff Burgdork, Jaak discovered that rats can laugh. Well it's more of a chirp, but humans can't hear it. The sound is emitted at 50 kilohertz—a pitch that is above the range for people to hear it.

How did Jaak make the rats laugh? He tickled them. It was an experiment to understand the nature of laughter, play, and joy. In a 2012 interview in *Discover Magazine* called "The Man who Makes Rats Laugh" Jaak said, "Joy is social, so you're looking at play. Play is a brain process that feels good, that allows the animal to engage fully with another animal. And if you understand the joy of play, I think you have the foundation of the nature of joy in general."

Jaak has been working on an idea of what he calls "emotional primes." In the interview he went on to say, "These are emotional primes, the primary-process emotional systems associated with specific brain networks and specifically designated in the brain-stimulation studies of emotion. They are SEEKING, RAGE, FEAR, LUST, CARE, PANIC/GRIEF and PLAY."

So for Jaak and his researchers, play and laughter are primary emotions for not only humans but many animals. But Jaak also points out something else that is interesting when you combine it with the idea that people want their devices to have this sense of humor and play. Jaak says that play and joy and inherently *social*. It's about the interaction between two people. Humor is social.

Humor Is Social

> *When humor goes, there goes civilization.*
>
> — **ERMA BOMBECK**

> *If you crack a joke in the woods and no one hears it, then is it really funny?*

Tragedy is when I fall into a hole, but comedy is me watching *you* fall in a hole. A joke has to be told *to* someone. A stand-up comedian needs an audience. What do all of these have in common? People. Comedy needs an audience. Humor is at its core about communication between people.

Mitch Earleywine got this early on in his 2010 book *Humor 101*. He wrote, "Comedy is inherently interactive. If an oak falls in the forest and no one sees it, it's not funny. If one person sees it, it might be. If two people see it, there's an even better chance that they'll laugh. The words *social* and *amusing* relate in intriguing ways. ... The presence of others, or even the sound of their chuckles, increases our own amusement."

Humor needs people. That has really interesting implications when we think about giving our technology and devices a sense of humor. Really we are giving them a way to relate to us. This is very much what Jamie said and Dr. Genevieve Bell and I talk about in the next chapter. If a technology can *get* humor then it can *get* humans. It can understand us in ways that will make our lives not only better but easier as well. To understand humor is to understand the nuances and culture of being human. Humor is cultural and has context. What's funny at the office is not the same thing as what's funny at the local pub. And this is okay. Just like people are complex, so too is humor.

I think that's really it. Steampunk and maker culture expose a broader human desire for technology to understand the subtleties of who we are. But there isn't just one of us. Just like there are many different types of humor from satire to bathroom jokes, so too are there varied and changing versions of ourselves.

The requirement of humor is the requirement to understand the complexity of who we are personally. That's a daunting requirement and a really interesting design challenge. We can approach the subtleties and complexities of people all over the world through the lens of humor. This gives us a place to start; a way to approach the problem. It's our first requirement: Just make'em laugh.

> *Among those whom I like, I can find no common denominator: but among those I love, I can: all of them make me laugh.*
>
> **— W. H. AUDEN**

Why Now?

In their April 21-27, 2012 issue, *The Economist* heralded a 14-page special report on The Third Industrial Revolution.

> *The first industrial revolution began in Britain in the late 18th century, with the mechanization of the textile industry. Tasks previously done laboriously by hand in hundreds of weaver's cottages were brought together in a single cotton mill, and the factory was born. The second industrial revolution came in the early 20th century, when Henry Ford mastered the moving assembly line and ushered in the age of mass production. The first two industrial revolutions made people richer and more productive. Now a third revolution is under way. Manufacturing is going digital. Like all revolutions, this one will be disruptive. Digital technology has already rocked the media and retailing industries, just as cotton mills crushed hand looms and the Model T put farriers out of work. Many will look at the factory of the future and shudder.*
>
> **– THE ECONOMIST**
>
> April 2012

The special report goes on to talk about how technological advances like 3D printing, collaborative manufacturing, and automation will revolutionize factories and the lives of people all over the planet. It's fascinating and terrifying at the same time.

Humor is a way for people to feel more comfortable with technology. Not only that but it also gives us a way to understand socially what's acceptable. Humor allows us to process passive change and get us talking about what's acceptable and what is forbidden. Humor doesn't answer these questions. It gives us an acceptable way to talk about it.

By laughing at a new technology or even at the massive changes of the Third Industrial Revolution we feel as if it's not that big of a deal. If a technology is the butt of a joke then it can't be that scary. If you can laugh at the thing that terrifies you then it's not that scary anymore. Plus, making fun of something also humanizes it. We know that your computer doesn't feel embarrassed when we make fun of it, but we collectively make it more human. Making fun of it gives us a way to start talking about it and we start to understand what it means to us and how we want it to fit into our lives.

Then once we have accepted the technology into our lives it opens up new spaces for humor and comedy. Like Jamie said: laughter and humor bring us closer not only to people but also to the devices and technology that bring us that joy.

The question then is: As we stand poised at the beginning of a third Industrial Revolution, what role will humor play? Steampunks, hackers, and makers are telling us that people and popular culture want their technology to have a sense of humor. But what specifically can it *do*?

Looking back at the previous two industrial revolutions might help. In Chapter 14, Dr. Genevieve Bell points out something interesting. She saw the steampunks as the romantic poets of the Information Age. The romantic poets were reacting against the first industrial revolution, using their art to explore their place in a technologically advancing world (we'll see more of this in the next chapter). Perhaps we can see the steampunks doing something similar with their work. The entire culture is searching for its place out in the forefront and soon the things they learn, the new narratives they are telling themselves about technology, will trickle into the mainstream.

Were there any examples of this from the Second Industrial Revolution that took place in America at the beginning of the 20th century? I did a little digging and out popped someone I wasn't really expecting: Charlie Chaplin. (It shouldn't have surprised me that much. Cory Doctorow had talked about Chaplin and *Modern Times* in Chapter 3.)

In 1936 Charlie wrote, produced, directed, and starred in *Modern Times*. (Chaplin actually scored the movies it as well... the guy had talent!) The film is a direct reaction and comment on the Great Depression and the Second Industrial Revolution.

Modern Times was a massive global success, proving that Charlie and the Tramp still had it. But why was it so popular? Audiences loved the images of Charlie raging against the machine, literally being eaten up by it. But why? The experts are divided.

You can't get any more mainstream than Leonard Maltin. Before the Internet, Leonard Maltin *was* the Internet for movies. If you wanted to know anything about a movie before 1995, you went to Leonard's book *Leonard Maltin's TV Movie and Video Guide*. Maltin is a trusted and wildly popular entertainment reviewer. He's so popular that he's had a guest appearance on *The Simpsons* and played himself in *Gremlins 2: The New Batch*.

Leonard summarizes *Modern Times* like this in the well-used 1990 version of his book that I had on my bookshelf:

> *Charlie attacks the machine age in unforgettable fashion, with sharp pokes at other social ills and the struggle for modern day survival... Chaplin's last silent film is consistently hilarious, and unforgettable. Final shot is among Chaplin's most famous, and most poignant.*
>
> **— LEONARD MALTIN**
>
> *Leonard Maltin's TV Movies and Video Guide* (1989)

Like most people, Maltin believes that Charlie is poking fun at the machine age, pointing out the troubles and personifying the audience's struggles. And Leonard is not alone. In his intense scholarly work *Machine-Age Comedy*, Michael North describes the effect of *Modern Times* this way:

> *The majority of city dwellers, throughout the workday in offices and factories, have to relinquish their humanity in the face of an apparatus. In the evening these same masses fill the cinemas, to witness the film actor taking revenge on their behalf not only by asserting his humanity (or what appears to them as such) against the apparatus, but by placing that apparatus in the service of his triumph.*
>
> **— MICHAEL NORTH**
>
> *Machine-Age Comedy: Modernist Literature and Culture*

So Charlie is raging against the machine, plain and simple. Well, maybe not. Paul Rotha has a very different view of *Modern Times* and what Chaplin might be saying. Paul, an Englishman, is the author of the massive 830 page book *The Film Til Now* (the hardback is nearly five inches thick). Paul started working on the book in 1929. He was 22-year-old aspiring filmmaker who couldn't get work in the English film business, so he took the assignment to write a book about film. In the preface to the book he says that he did it as a way to break into the business.

Unfortunately, after the release of the book, Paul found it even harder to get work until he finally broke into the industry with documentary films. With the help of a friend of his, Richard Griffith, an American, the two updated and expanded the book over the years. Their view of *Modern Times* went like this:

> *This one man rebellion of Charlie's was tolerated by social optimists of all parties because it seemed the last expiring effort of romantic individualists. But when, in Modern Times, Charlie specified the inimical society to be an industrial one, all shades of opinion were discounted. Chaplin had left Chaplinland and ventured into the real world. His very presence was a criticism of the world, but a criticism that led nowhere that anybody wanted to go. He presented no remedy. He did not stigmatize one faction in order to uphold another; he stigmatized them all. He seemed an anarchist. He was as unready to save the world for this as for that. Socio-economic prescriptions for socio-economic ills were none of his. No one could take comfort in him.*
>
> *It is my contention that he does not want, and never has wanted anyone to take comfort in him. Comfort, despite the annealing power of laughter, is not his line of goods. He is a satirist and satire has no allegiances.*
>
> — **PAUL ROTHA**
>
> *The Film Till Now*

So which is it? Is Chaplin poking fun at the ills of the second industrial revolution—making a judgment on the dehumanization of the modern worker—or is he offering no answers at all? Charlie was exploring the role of humanity and technology. The Tramp is literally swallowed up by Henry Ford's machine. Powerful stuff for people who were in the depth of the depression, who had grown up on farms and a very different America and now found themselves working in a factory if they were lucky. The Tramp simply becomes a proxy for the audience. At the end of the movie Chaplin gives no answers. Funnily, he ends on a note of optimism and the Tramp and the ingénue walk off into the sun. But Chaplin the writer and director gives no answers, he simply expresses and explores the public sentiment.

Like the steampunks exploring this 21st century revolution, they may not have answers but it's in the exploration that we will find our way. There were no answers in Charlie's movie house but it was the comedy, the laughing, that allowed a kind of mass coping mechanism for society and culture; maybe not to understand, but possibly just to think through what was happening to them in their daylight hours. The Tramp was the vehicle, just as steampunk is the vehicle to process what is happening now.

Accepting the New

Humor isn't how we find the answers, but it's more of a process. Like James said at the dinner back in Chapter 5: "culture change is messy."

So why do people want their technology to have a sense of humor?

Humor and laughter connect us to the technology. If a technology can help you laugh then you don't see it as technology anymore. If it makes you laugh, gives you comfort, then you are connected to it, you actually see it as a conduit to the people and things you love. You stop seeing it as a cold device; you see it as the things the device gives you and so then you accept the technology because it gives you or connects you to the things you want.

Humor is how we learn to accept the new. It's the broad cultural process by which we understand what is new in our culture. Humor we get, humor we understand. Humor allows us to feel more human with the thing that is not human but is rapidly becoming a part of our daily lives. Humor not only allows us to process the new technology but it allows us to work out the cultural rules around that technology and how it fits into our broader culture.

> *Make 'em laugh... Make 'em laugh. Don't you know everyone wants to laugh?*
>
> **— NACIO HERB BROWN AND ARTHUR FREED**
>
> "Singin' in the Rain" (1952)

14

Don't Forget the Humans

Humor, History, and Humanity—these are all intensely human ways of thinking about technology. Is it odd that so many people are looking for their devices and gadgets to be more human? Why? What does it mean to make technology more human? And can we even do it?

To the scientists!

On a call between America and Australia, Brian talks with world-renowned cultural anthropologist Dr. Genevieve Bell about the relationship between technology and people. In her trademark scholarly and surprisingly irreverent style she explains how throughout history humans have always searched for ways for our tools and technologies around us to adapt to our needs. Not only that but we use humor and history as a way to allow these technologies into our lives.

How an Anthropologist Changed the World: Genevieve Bell

Brian David Johnson (Portland, Oregon)

There's a joke in Genevieve Bell's family that anthropology is not a career, it's a life choice. Genevieve is the daughter of an anthropologist. Her mother Diane Bell studied anthropology after she had begun to raise Genevieve and her brother. In her 20s Diane attended university on a single mother's budget, which was next to nothing. When it was time for her to go to her first anthropology class she only had enough money to put one of her kids in day care. So while her younger brother was sent off to play with the other kids, four and a half year old Genevieve went to Isobel Wright's Introduction to Social Organization class with her mother.

Sitting under Diane's desk, Genevieve listened as Professor Wright explained multilateral cross-kinship marriage or a preferred marriage pattern. Always a blisteringly fast learner, young Genevieve worked out the system better and quicker than anyone else in the class. Never one to keep her opinions to herself, the four and a half year old got up from under the desk and explained to the rest of the class that it was a way for you to marry your mother's mother's brother's children's children.

This didn't make the class of Melbourne college students very happy. Who wants to be shown up by a four and a half year old on their first day of anthropology class? So Genevieve was sent out into the hall with her coloring books and toys to wait out the rest of the lecture. (I love the image of a young Genevieve coloring on the floor of the university halls, too smart to be let back into class. Her mother likes to joke that Genevieve was kicked out of her first anthropology class.) So it's pretty easy to see that Genevieve was born and raised to understand humans very quickly.

But anthropology and the science of people was not the direct path for Genevieve. It wasn't until she found herself in the United States years later that she found her true calling. She was attending Bryn Mawr college in Pennsylvania and had just finished up most of her undergrad studies when she wandered into the Anthropology department. It was a life-changing moment for her. A died-in-the-wool Aussie, Genevieve loves Australia and still spends much of the year there. She says that when she wandered into the Anthropology department at Bryn Mawr it was the first time in years she didn't miss Australia. She was 13,000 miles away from home and she didn't feel homesick. Listening and understanding people was home to her.

In 1998 Genevieve left a promising career in academia; she was on the tenure track at Stanford University in California when she accepted a position at Intel. I was fascinated to find out how a born social scientist ends up at one of the largest high tech companies in the world. How does a cultural anthropologist end up at an American manufacturing company that makes silicon in factories all around the world? The answer is a little complicated.

Anthropologists were a new thing in corporate culture. One way to think about anthropology's gradual adoption into industry is as separated into three distinct waves. The first wave were consultants—think Mad Men and the glitzy Madison Avenue advertising agencies of the '50s and '60s. These were social scientists studying human behavior. The Madison Avenue agencies needed to understand people so that they could sell to them. These Mad Men didn't care where the information came from; if it gave their product a boost in sales then they would take it. Call the social scientists! What does the common housewife want? Can buying one laundry detergent over another mean that you love your family more? What does the young executive want to smoke and drink? (Remember this was the '50s.) Can a car make you more masculine? Can a cigarette make you more worldly and sophisticated? Consumer culture was king, and the pitchmen on behalf of their companies wanted to understand the minds of those consumers.

The second wave found these social scientists bravely embedded in the companies themselves. No longer experts or consultants, they were a part of the machine. Most were relegated to the marketing departments. Again, companies wanted to understand their audiences so they could sell to them better products. No longer was it just about the message or the advertising pitch, now it was about understanding the minds of the consumers so that these companies could *make* better products for them.

In the '70s and '80s there were a handful of luminaries that made the jump from the classroom to the boardroom. Applied-anthropology legends like Lucy Suchman and Julian Orr worked in places like the legendary California Xerox PARC lab or on the east coast of America with companies like GM and large insurance agencies. All used their scientific approach to understanding people and applied it to the development of products and services.

The third wave found a very different place in companies; research and development departments (or R&D as they were known). It's not surprising that many of these R&D departments were in technology companies. The latter half of the 20th century saw an explosion of products and services that were enabled by computers, the microchip, and the Internet. Probably one of the most famous and historic examples of social science's influence on corporate American and product design is Xerox PARC.

As with most amazing moments in technology and innovation PARC was lightning in a bottle. Smart people at the right company at the right point in time. BAM! Magic. Xerox was looking for ethnography and anthropology to do what it was really good at: telling stories, making sense of the world, and carrying the voices of people who aren't in the conversation into the conversation. They were exploring what all these new and amazing technologies could do for people. It was a heady and amazing time.

This was the idea that brought Genevieve to the Intel Corporation. She had a driving passion for taking her expertise and having a large impact on people and technology. For her, social science was about changing how people look at the world. She literally used science to broaden and deepen people's understanding of other people. She was also committed to not dumbing down the science of the work. She wasn't just interested in making an impact at a global company like Intel, she also wanted to change how Intel and the entire technology industry designed products for people. If she did it right, that payoff would be huge. The result would change the lives of people all around the world.

In 2001, Genevieve embarked on a groundbreaking study called "Inside Asia." At the time, China was not the China that we know today. India was not today's India. With a remarkable vision Genevieve set off on a two-year intense research project traveling across India, China, Korea, Indonesia, Australia, Malaysia, and Singapore. She shopped with families in Pune, India. She went to temple in Pusan, South Korea surrounded by baby Buddahs. She sat down to eat with a family in Ipoh, Malaysia to have some of the best noodles in her life.

The goal of the research was to understand Asia not as Asia—not as the pundits and the economists understood Asia; they have always seen them as a collection of numbers, gross domestic product, population, and statistics—but to understand Asia as Southeast Asia, India, and China. To understand the people, how they lived, and who they were so that Intel could build better products for them.

At the time people would say things like: "India is the next China." But Genevieve was bold enough to disagree. Granted they both had more than a billion people, but that's about where the similarity ended. India and China have very different cultural behaviors and very different ways of adopting technology into their lives.

Back then the response was typically, "Yeah, yeah, yeah... but once India and China get enough money they will be just like the USA." In recent years this has shown to be false. Genevieve showed that each country had wildly different anxieties and regulations that needed to be understood and honored for any company to really appeal to these markets.

I've known and worked with Genevieve for nearly a decade. She told me once, "To make real change at a large company it's not good enough to just do research and come back and tell the stories. Don't get me wrong, that's incredibly important. You have to do good solid social science. But you can't stop there. You also have to change the company. You have to change the process by which the technology and the products are developed. That's how you make lasting change. That's how you make it stick."

Genevieve's work could not have been more accurate. Since then, Intel's profit center for its platforms has diversified from the US and Western Europe to China. Asia has become one of the company's largest markets. Success and her efforts have been nothing less than transformative. In 2012, CEO Paul Otellini announced that Intel was no longer just a chip company or a high tech company, they were now an experience company.

Genevieve has since been awarded the title of Fellow, one of the highest ranks at the company. Not bad for a social scientist from the outback of Australia. Genevieve has come a long way from sitting under her mother's desk and getting kicked out of her first anthropology class at age 4. But her passion, intelligence, and furious commitment to the human perspective has transformed the high tech industry and touched the lives of billions of people.

What better person to sit down and talk with about what steampunk and maker cultures have to say about the evolving relationship with people and technology? Now, I've worked with Genevieve for years, so I have a pretty close understanding of her world. But for this conversation we were both on opposite sides of the world. I was in Portland, Oregon while she was in Bryon Bay, Australia.

Dr. Bell's view in Australia during our call—that's her "flat white" on the window sill

A Genealogy of Devices: Humor Is All About Relationships

If you have never been to Australia, one of the unexpected things you notice is that the birds are really loud. No, really, the birds are *crazy* loud. I'd forgotten about this until Genevieve and I stared chatting. The entire time we talked there was an outrageous cacophony of squeaks, squawks, and ear-splitting calls that bordered on the ridiculous. Still we persisted.

"What did it mean that people wanted their technologies to have a greater sense of humor?" I asked.

"That's really an interesting idea," Genevieve replied. "Humor is one of those profoundly human and personal things. People's humor and sense of humor differs all over the world. One of the things that's fun about technology and humor is that for some time now technology has been a great source.

"Giving technology a sense of humor would be radically different than what we're doing today," she continued. "Today in technology we are all about command and control. We tell the technology what we want and we want the technology to do it. Supposedly the better the technology does what we ask, the better we will like it. But giving technology a sense of humor means that we need to start having a relationship with it... because humor is all about relationships."

"What do you mean?" I jumped in.

"There are some early examples of that today..." she paused and thought for a moment. "Every now and again you see these break-out moments when technology actually engages with us. Perfect example: I like the GPS devices that can be programmed to have different voices like Darth Vader and Snoop Dogg."

"That's at least the beginning of a relationship," Genevieve continued. "You as a driver will have a relationship with your GPS, if Vader or Snoop are talking to you, telling you to turn left at the next intersection. The device has now taken on the personality of a well-known star that you know and like. And come on, having Darth Vader and Snoop Dogg give you directions is kind of funny. But this is all early days.

"I can imagine a time where all our devices do have personalities," she continued. "They will have to have personalities if we want to have a relationship with them and that kind of interaction takes a lot of intelligence. The device will have to know you pretty well if it's going to be funny or at least a little bit cheeky. But once you have that much intelligence in the device then you move to a whole new level of interaction. The device now can start to anticipate what you might want. If your device can make you laugh then it knows you well enough to make decisions and assumptions about what you might like. This is far more elegant than the command and control approach we take today. Giving a device a sense of humor would really mean making it far more intelligent and frankly a much more interesting device to interact with."

"Do you think some people will get worried if the device gets too intelligent?" I asked.

"We are at a really interesting point in time," she started. "You're right, people do have those fears but it's a very Western fear. There's a long history of the western world having an anxiety about making our technology too intelligent. The anxiety is that if we make computers or technology too smart then it will somehow get the upper hand and if they get the upper hand they will do away with humans." She laughed. "This says so much more about humans and what we think of ourselves than intelligent technology.

"I think this fear is us working out our fears around what it will mean to live in a world where we have devices and technology that are smarter than us," she continued. "From a technology stand-point we are clearly getting there. Artificial Intelligence lands our planes and parks our cars; it won't be long before it's doing a lot more than that.

"This is where I think humor really becomes important," she paused. "Humor is how people start to become comfortable with new technologies. The jokes we make about the machines we use happens earlier and more often than us worrying about the machines killing us."

"What do you mean?" I asked.

"Look at mobile phones," she explained. "We live in a world where mobile phones are in our hands, in our pockets, and certainly close to us all the time. Think about all the jokes about your mobile phone. Stupid things people do with phones. Funny ring tones. These are all new behaviors and situations we find ourselves in. Is it ok to talk on a phone in a bus? In the bathroom? In church? We are starting to incorporate them into our daily rituals and at the same time we're seeing cultural push back. This is really the pattern technology goes through as it moves from being something in the labs to something in our lives.

"I remember being in India in 2002, when I was doing field work, and people teasing each other about new flashy mobile phones. They would say: 'Don't be so Punjabi about that!' Meaning don't be so bold and ostentatious. In this case it wasn't about the devices being funny, it was more about the way people use teasing and humor to naturalize devices into our lives. People all over the world use humor to take the steam out of technology, to make it easier to deal with."

"As we were finishing up the research for Vintage Tomorrows, the iPhone 4 came out with Siri," I started. "James and I found it interesting and very telling that when people talked about Siri or wrote reviews they usually wouldn't mention how great it was at finding you a good place to get sushi. What everyone was talking and writing about were the jokes. Siri had a sense of humor. There was even a YouTube video of Siri talking to a Furby."

"That was one of the funniest little bits of video that turned up on the internet," Genevieve laughed. "Watching Siri, this amazing new technology talk to Furby, this once-amazing technology that was now just funny was really telling for me. I think there was something going on underneath that interaction."

"What do you mean?" I asked.

"Well it was more than just funny," she replied. "Having Siri and Furby talk together placed Siri in a genealogy of devices. It gave people a way to place Siri in a timeline of other somewhat intelligent devices and it used humor to do it. It was an example of people using humor for us to get comfortable with what could have been seen as a really scary, too-intelligent technology."

History Is a Tricky Idea

"Funny you should mention using humor to give a device a sense of history," I jumped in. "That was the second idea that has emerged from our research. We did see that people wanted their devices to have a sense of history."

"Well yes, of course," Genevieve scoffed. (Yes, she actually scoffed.) "As an anthropologist I would say that a sense of history is what makes us human.

"But history is a tricky idea," she continued. "People will always want a sense of history, a sense of where they are in time. But they will also always want to reinvent themselves. They don't want to lose their history but they also don't want to be welded to it. That's fine for historians but it's tricky when you try to apply it to technology. From a device standpoint, that would mean striking a balance between a device knowing where you've been and what you've been doing and remembering how you like to do things, but at the same time doing this without it feeling like your mother is going through your sock drawer."

"What?" I couldn't help but stop her.

"That's what my mother used to say," she laughed. "My mum always used to say that people want a history but they also don't want to feel like people are rummaging through their most secret places. People want to be who they are and have a sense of history but they also want to keep their secrets and have the ability to lie a little about themselves."

"That makes sense," I added.

"I love that image because it's really about how people balance history with the desire not to have everything out in the open," she paused. "People keep secrets. There's a delicate balance between remembering what we want to remember and remembering what we need to remember."

"Personal histories and public histories?" I asked.

"They are the same things," Genevieve answered. "But we don't have to make it as grand as that. How about we design technologies that are polite and remember they've met you before!" She laughed. "I was talking to a someone recently who told me how incredibly frustrating it was to him that every time he turned on his Nintendo Wii the device would reintroduce itself to him. It was like they had never met before. That's just rude." She laughed again. "Imagine if I did that as a person." Still laughing.

"Ok..." I filled the space in the laughter.

"For example," she composed herself. "I've known you, Brian David Johnson, for ten years. But imagine that every time I saw you I said, 'Hello there, I'm Dr. Genevieve Bell! It's nice to meet you. What's your name?'" She chuckled and sounded like she was wiping the tears from her eyes. "On a human level it's just plain rude, but our devices do it all the time."

I jumped in here. "James, our cultural historian, likes to ask people, 'What kind of history do you want to be from?' He says that the past is always in motion and that the steampunks are remaking their past, bringing people into the past who previously didn't have a voice."

"I think that's fun," she answered. "People say a lot through the history they choose to be from. It's telling what history people write down. It's a statement of who they want to be and also who they don't want to be. It's always important to understand who's telling the history. But it's just as important to understand who is editing the history. Everything doesn't always make it into the record.

"Memory isn't perfect and it's always going to be a little fungible. But history has a definite power," she finished up. "It's the power for how things are remembered and who gets to do the remembering. It tells you who is in charge and very distinctly what they want. That's why having people rewrite history is really interesting. It's why steampunk is interesting."

Infusing Human Personality Into Technology

"The last big area that we saw was that people wanted their devices to have more of a sense of humanity." I started.

"Definitely," Genevieve replied quickly. "For years in computer science, we have imagined that humanity equals intellect rather than consciousness. We've talked about artificial *intelligence*, not artificial affect or artificial anything else. The human stuff, the stuff about personality or humanity stuff has been left for science fiction to explore. Just think of all the jokes in science fiction about objects with personalities. I love the K-9 robot in the *Doctor Who* series. It was just belligerent. And also the jokes in Douglas Adams's science fiction. I love the self-satisfied doors that sigh when they open and close. That's personality, and it's funny," she laughed.

> *"Ghastly," continued Marvin, "it all is. Absolutely ghastly. Just don't even talk about it. Look at this door," he said, stepping through it. The irony circuits cut into his voice modulator as he mimicked the style of the sales brochure. "All the doors in this spaceship have a cheerful and sunny disposition. It is their pleasure to open for you, and their satisfaction to close again with the knowledge of a job well done."*
>
> **— DOUGLAS ADAMS**
>
> *The Hitchhiker's Guide to the Galaxy* (1979)

"All of this science fiction is about technology getting infused with personality, with humanity," Genevieve continued. "It's happening more and more. Over the last 15 years, robots and computers in science fiction TV have the ability to talk back. That's interesting and something that you see in science fiction but not in science fact yet..." she paused. "What this is telling us is that our relationship with the technology around us is not just about compliance. Sometimes you don't need the robot or technology to talk back to you but we still have it respond. This means that we obviously want some kind of engagement. People want their technology to talk back to them so we can have a relationship with it. We want it so much that we put it into all our fiction. That sense of humanity and interaction is important to us.

"But in the world of technology development and engineering we're not thinking about it in this way. The best example of this is the Turing Test.

"The Turing Test has set up a very particular type of engagement," Genevieve explained. "The interaction that it described is how we imagine our interaction with machines should be. It's really testing can a human being tell the difference between the answer of the computer and the answer of the person. But this is all wrong. These are questions about rationality and logic. They're not questions about humanity or feelings or relationships. The central idea in technology development and the Turing Test is can a machine think like a person. Not can a machine feel like a person."

"I see," I said.

"It would be very different to test for feelings. Or test for being in a relationship? This idea strikes really deep into who we are. It makes us ask what makes humans human."

"And..." Remember those Australian birds I told you about at the beginning of the chapter? Well, they were really loud now. I strained to hear what Genevieve was going to say.

"From a logical social science point of view, humans are nested in relationships; we are a part of systems and relationships. That's a very different way to look a humanity and technology."

The Romantic Poets of the Information Age

"What is your take on steampunks and hackers," I asked, wrapping up the conversation. "What do you think about this world that we've been talking about?"

"I'm torn about it, truthfully," she said flatly. "It's a powerful form of historical revision. Anyone who engages in the revision of history is engaging in an interesting and dangerous set of activities. But because this is so powerful you have to be careful that it's done in an informed way. I don't see a lot of the steampunks and hackers talking about real history. How do we retell the present by retelling the past? It's notably absent from most of what I see coming out of hackers and makers. I'm worried that some run the risk of having a deep affect without thinking about the consequences. It can come off as a kind of privileged activity.

"But with all of that said," she slowed down to make the point. "Like many forms of social critique, this is really powerful. It's captured people's imaginations. It does have something to say. It resonates by remaking technology and suggesting a different trajectory for technology. It taps into our current anxieties and the desire for a sense of wonder. There are fleeting moments today of magic and wonder, but they're far less than they used to be. I think steampunk and hackers offer a moment of wonderment, and that's a powerful and important thing. I've been thinking about steampunks as the romantic poets for the information age.

"The romantic poets, like the steampunks, were engaged in a revision of the historical projects of their day," Genevieve explained. "The romantic poets like Lord Byron and Percy Shelley found themselves in a time with little romance or wonder. It was the first industrial revolution and technology was fundamentally changing the physical landscape around them as well as their entire social order. Industrialization was definitely landing on British shoulders and changing daily life fundamentally. But it wasn't all bad. There were ways that they were taking advantage of the changes. Communication over long distances was better. They had mail delivered to Geneva eight times a day—that sounds almost like email!" she laughed. "Books were getting cheaper and travel was more accessible. It was a mix of good and bad, a mix of the dark satanic mills and a new industrialized age.

"The romantic poets were attempting to reclaim an earlier simpler time period, but it was a time in early Britain that ever happened. They were attempting to look back to early British moments, and then think about reclaiming them for their present. The makers, hackers, and steampunks are doing the same thing. They are trying to offer additional perspectives. They are providing a form of cultural critique, like romantic poets. And like romantic poets they will have a powerful influence on culture and the future far beyond their numbers. It is a form of social critique that's very popular. You have Justin Bieber and Nicky Minaj making videos in the motif—that shows you how popular it is and how deep a chord it is striking with people."

15

When Is an iPhone Like a Pocket Watch?

The entire *Vintage Tomorrows* gang heads to New York City to attend New York Comic Con. Byrd is doing a preview screening of the documentary while James and Brian finish up on their research and talk about the importance of history and technology.

History anchors our understanding of the new; it gives our lives resonance and allows us to make sense of an increasingly complex world. James reminds us that many futures all happening at once require an equal number of simultaneous pasts. Just like the future, history lives in our imaginations. What does this mean for how we build new technology? How do we make different requests of the future, and what kind of past do we want to have? Brian explores the science of memory and future visions, showing us that history is not only the onramp to the future but that they are actually two sign posts on the same road.

Is That the Past in Your Pocket?

Brian David Johnson and James H. Carrott (New York City, New York)

We were nearing the end of the project and we met up at a bar in New York City. We were in town for Comic Con NYC, talking about *Vintage Tomorrows*. Byrd MacDonald and the Porter Panther crew had done a sneak preview screening of his documentary and it went great. Even more amazing was the panel that followed. Lined up in a long row we had Byrd, James, Brian, Anina Bennett and Paul Guinan, Cory Doctorow, and Alan, the editor of the movie. The conversation was knock-down drag-out awesome.

Brian and James were pretty jazzed. We took the time to get together and catch up on what we had learned about one of our "big three": Humor. History. Humanity.

Brian figured that James, the cultural historian, would have this one nailed. But James was distant.

"Where'd you go, dude?" Brian asked. "You got real quiet all of a sudden."

Worth noting, James is not generally the quiet type. "It's this darn thing." He pulled his iPhone out of his pocket and set it on the table.

"Which thing?" Brian was still looking through the beer menu, pondering the eternal question: Which IPA to sip alongside the shot of whiskey. It had become a habit for these meetings.

"This one," James pointed. "The shiny black thing that's on the table and isn't a coaster."

"Ah." Brian glanced away from the menu then back again. "What about it?"

"It's the embodiment of *not steampunk*," James said thoughtfully. "But we've all got them in our pockets. Most of the steampunks we talk to have smartphones. But it's so not steampunk."

"Well, what would be more steampunk?" Brian asked wryly.

"Pocket telegraph?" James said quickly. "Hell, I dunno. But not this." James pulled up the "Steampunk Rotary Dial" app he'd downloaded for $0.99 a few weeks before. "I mean, who doesn't love a rotary phone?" Brian was not amused. "But this isn't steampunk at all. It just pretends to show the phone with its cover off."

Brian chuckled. "Okay, so let's play with this idea a bit. What do the steampunks have to say?"

"Well, we always end up talking about pocket watches," James admits.

"So let's talk about pocket watches," Brian prodded James.

"You know me," James said, knocking back the last of his scotch, "I'm always happy to talk pocket watches, but I'm trying to get at something bigger."

Brian leaned knowingly forward. "Indulge me," he said, waving to the bartender to order.

How could James refuse?

Once Upon a Time, the Pocket Watch Was an iPhone

James H. Carrott

The iPhone is the anti-steampunk. It is a piece of glass, slick and shiny. It almost disappears when switched off, and becomes indistinguishable from any other iPhone (okay, some of them are white... you get the point). Apple's approach to design has become extremely influential, but it's hardly alone in its drive for sleek simplicity. Physical technology today is designed to disappear into the background of our lives.

The iSlate

As you might imagine (having read Chapter 10) market-vigilant minds at Mark Thomson's Institute of Backyard Studies have an entry of their own in this conversation: the iSlate. I was fortunate enough to be invited to their leading development shed for a sneak peek of this hot new product.

Featuring a convenient, wipeable real Australian slate "slate" and a handy chalk compartment, the iSlate is a clear contender for consumers who prize simplicity over all those fancy apps. My brief investigation did not reveal the purpose behind the team's focused fruit consumption. I suspect a clever marketing ploy.

As brilliant as it is, however, the iSlate did not solve my conundrum. Probably because it is *too* simple. Chalkpunk may be ahead of its time.

In the 21st century the pocket watch is rarely more than a symbol or a fashion accessory. It is the opposite of the iPhone. It's a symbol for history and a time that has passed. When you think about it, why would anyone carry a pocket watch in the 21st century? The ability to find out the time is spread across all the devices and technology in our lives. But a pocket watch still means something.

iSlate 2 and iSlate 3 prototypes

In its day, the pocket watch *was* an iPhone. It was an amazing device that put time itself in your pocket. It connected its bearer to a broader synchronous world, increasingly stitched together by wire, rail, and steam. Empowering, sure, but it also produced a bit of anxiety. After all, Alice, who carried no watch of her own, was simply lying about on a hot day, lazily considering daisies...

> *...when suddenly a White Rabbit with pink eyes ran close by her.*
>
> *There was nothing so very remarkable in that; nor did Alice think it so very much out of the way to hear the Rabbit say to itself "Oh dear! Oh dear! I shall be too late!" (when she thought it afterwards it occurred to her that she ought to have wondered at this, but at the time it all seemed quite natural); but, then the Rabbit actually* took a watch out of its waistcoat-pocket, *and looked at it, then hurried on, Alice started to her feet...*
>
> **— LEWIS CARROLL**
>
> *Alice's Adventures in Wonderland* (1865)

What makes the smartphone wildly different than the humble pocket watch is its ability to adapt and customize meet a variety of needs. The function of a pocket watch is crystal clear. It tells time. But an iPhone (or Android phone, or whatever ___ phone comes next) has apps. When you fire up your smartphone you could be doing almost anything... on a device that is, from a design point of view, trying to disappear. We've shifted our lens from hardware to software, creating a hidden virtual world behind the device.

Broadly speaking, steampunks and contemporary makers operate in reaction to this, arguing that our technology should be open and visible. It wasn't the White Rabbit that first caught Alice's attention, but watch. She'd have missed the adventure of a lifetime if that bunny tech was invisible. The sentimental object reigns supreme. This idea is so accepted in these cultures that even talking about it has become a cliché.

There was a time when you could take apart devices. A pocket watch is just one example. People took apart things like rotary phones, transistor radios and cigarette lighters. An ordinary person could take one of these apart, understand how it worked, and maybe even put it back together again! Empowering, right? You were smarter than the device. You understood how it worked.

When this topic came up on a Norwescon 2011 panel, Pyr Books publisher Lou Anders chimed in with a slightly different take: "I'm sure I couldn't put that pocket watch back together again. Technology has always been at a remove from the people." His point was that the whole pocket watch thing is symbolic. We might not actually be able to take any older device apart and put it back together, but steampunk technology makes us *feel* like we can. It gives us that sense of understanding—of empowerment; the feeling that we are the ones in control, not the machines. It isn't just the steampunks, either. We seem, as a culture, to be developing a broader feeling that technology is moving too fast, getting too smart, and we are losing control. The title of Thomas Thwaites's brilliant little book says it all: *The Toaster Project, or A Heroic Attempt to Build a Simple Electrical Appliance from Scratch* (2011). Even the mundane now feels beyond our grasp.

Meanwhile Back at the Bar

"I love it!" Brian shouted. "I love it a whole big bunch. So smartphones are the antithesis of steampunk technology. Because the technology inside the smartphone, the hardware and software, the services, and connection to the Internet are all invisible. Precisely what makes it cool and magic is unseen." He took a sip of beer. "This is only going to continue. But what steampunk is doing is trying to make those things visible... hence the gears."

"Right," James replied with a smile.

Brian was nearly jumping in his seat. "That's why steampunk is obsessed with gears."

"Absolutely." James continued to smile. "They turn things inside out. People look at steampunk culture from the outside and have the misperception that steampunk is anti-technological or backward-looking. But the truth is quite the opposite."

"But what they are doing with that technology, the relationship they are having with it is very different than mainstream tech." Brian paused. "By making all this magical technology into the Victorian era they are giving it a past."

"They are giving it a *new* past," James added. "It's a past they never had. But it's one they want. History gives us a way to understand the future."

We Can't Touch Our Past; It's Been Turned into Bits

James H. Carrott

> *Joshua Joseph has no great hatred of modern technology—he just mistrusts the effortless, textureless surfaces and the ease with which it trains you to do things in the way most convenient to the machine. Above all, he mistrusts duplication. A rare thing becomes a commonplace thing. A skill becomes a feature. The end is more important than the means. The child of the soul gives place to a product of the system. For anything really important, Joe prefers something with a history, an item which can name the hand which assembled it and will warm to the one who deploys it. A thing of life, rather than one of the many consumer items which use humans to make more clutter; strange parasitic devices with their own weird little ecosystems.*
>
> **— NICK HARKAWAY**
>
> *Angelmaker* (2012)

What function does the past serve in our lives? The relationships we have with objects has a lot to do with our history. Joshua Joseph Spork trusts things with a past because he can relate to them. When a thing has a past, when it carries history, it has a place in our lives. Its story becomes part of our story. It's a lot easier to trust an object you know than a sleek-looking stranger that passes in the night. My grandfather's pocket knife isn't just a blade, it's a part of him passed down to me.

We love what our devices do, but they're cultural blanks. They're empty. If you drop your smartphone in the toilet, it goes from magical communication node and life repository to useless piece of glass. But never fear! All you need to do is take an expensive trip to the store, plug the new one into your computer and within hours, it's exactly the same. We have no emotional attachment to this physical object. It's a means to an end. Hardly surprising that many of us find this alienating. The stories—the stuff that gives our stuff meaning—are digital now. They're in the ether and trapped behind glass. We can no longer touch our past; it's been turned into bits.

Steampunk shows us the element of human need in our technology. Even the people hot gluing gears on their iPhone are showing us that they want more from that device. It's physical. They want to make that faceless sealed device more personable and understandable. They want it to be physical—to have a tangible place in their history.

People want their technology to have a sense of history. We need that history to understand it and give it meaning in our lives. By creating artifacts of an imagined technological past, steampunk provides narrative roots for the future. Steampunk can throw our jet-lagged souls a rope, giving our culture a way to catch up with itself. Steampunk gives our future a past.

Yesterday is an essential part of today. Both are essential part of tomorrow. History tells us who we were, forming the foundation of who we are. "People feel like we don't have a history anymore in our culture," Jaymee Goh told me during our research. "But culture is the water that a fish lives in," she explained. "A fish doesn't know it's water until it's out of it and they start dying. That's what happens when you try to claim that you're cultureless—it means that you are dead. You are in it, you are living it, you are breathing it. And you situate what your culture is through history."

Steampunk uses history to make sense of the increasing complexity of the technology that surrounds us. It is a symptom of the broader cultural wrestling match between who we are and where we are headed.

MEANWHILE BACK AT THE BAR

"What you're saying is that all this playing around with the past is the symptom of something larger in our culture?" Brian asked.

"Yes!" James yelled back. The bar had gotten more crowded and the flurry of activity and voices was growing. "You keep making all these technologies smaller and faster and more invisible, so we have to use the past to understand them. People need the grounding."

"Well it isn't just me making the devices." Brian grinned and took a sip of beer.

"But you get what I mean, right?" James pushed his glasses back on his nose. "The past is essential to the future."

"Now that's saying something!" Brian lit up.

The Science of Memory and the Future

Brian David Johnson

As a futurist I'm always looking for new mental models, new ways that people are using to understand and use technologies. We talked with Bruce Mau about this earlier in Chapter 16. In the 1980s, to understand the personal computer we had to use the mental model of the page and books. Web pages, folders, and files were the metaphors that we used so that we could make sense and operate an incredibly complicated device. Without that metaphor, computers would have never gone mainstream. They would have remained imprisoned in universities and large government buildings, operated by a handful of people who could program them.

That's why mental models are so important and why I'm constantly looking for new examples. When it comes to the future of technology, if we can get the metaphor for how people will act and interact with technology right, then we can do some really amazing things.

James's idea that steampunk is a way culture is learning to deal with increasingly complicated technology fascinated me. Could we use the past as a kind of interface to our future? To understand this a bit more I did some research into the science of memory and the brain.

It turns out how we remember the past and envision the future actually happens in the same part of our brains. For decades neuroscientists weren't sure where memory "lived" in the brain but recent studies have shown that long term episodic memory comes from the hippocampus. The hippocampus is buried deep within our brains and is shaped almost exactly like a sea horse. This tiny part of the brain processes our memories and also our visions of the future.

In 2011 husband-and-wife neuroscientist team Edvard I. Moser and May-Britt Moser measured the brain activity in the hippocampus and showed that it "fired" during both "replay" (memory) of an event as well as "preplay" (future visioning). They observed the brain activity of rats as they made their way through a maze and also as the rats had a snack and a rest. During their break halfway through the maze the rats remembered making their way through the course (replay) but they also imagined what the rest of the maze might be like (preplay). The rats basically daydreamed a little and had visions of the future while munching away.

The remarkable thing about the Mosers' findings was that the hippocampus was being used for both processing memory as well as envisioning the future. Since then scientists around the world have zeroed in on this portion of the brain. In 2012 Matt Wilson, M.I.T. neuroscientist, was able to observe not only rats' memories and visions of the future but their dreams as well. He monitored rats as they dreamed about running through a maze. Then with the help of a little training and subtle audio cues he and his researchers could prompt the dreamers to change their paths through the imaginary course. Wilson used his knowledge of the hippocampus to control the rats' dreams!

What does all of this have to do with steampunk and the future of technology? How we process the past and how we imagine the future come both figuratively and literally from the same place. Long-term episodic memory allows humans to engage in evolutionary "mental time travel." This is a skill that has given us quite the advantage over the rest of the animal kingdom. We are not the only vertebrates that can imagine the future, we just happen to be really good at it. Steampunk's imagining and reimagining of our pasts and futures is an example of this mental time travel. But in this case the time machine goes forward *and* backward. Steampunk is not only remembering what happened, like the Mosers' rats in the maze, but it is "re-rememebering," revising the past to process the future.

Accomplishing these feats of mental gymnastics takes an extremely evolved brain. Even the human brain needs to progress and develop before it can engage in this kind of mental time travel. In humans, episodic memory usually doesn't kick in until we are about 2 or 3 years old and even then it's hazy. Try to remember your first memory. It's probably pretty spotty and incomplete. As we grow up the neural pathways in our brains that lead to our friend the hippocampus are just beginning to develop. But there's also another key factor that allows human brains to work at the complex level: our grasp of language.

Professor Martin Conway conducted language research at the City University London in the UK. In 2012 his work was explored in the October 6th edition of the *New Scientist* magazine. Conway's research showed that "children don't tend to remember an event until they have learned the words to describe it." We need the words to help us remember events in the long term. He went on to explain that words are the scaffolding upon which we hang our memories.

The mental models that steampunk is playing around with make even more sense when explored on the neurological level. Our understanding of the past and the future live in the same part of our brain. The act of remembering and the act of looking into the future are essentially the same action. On top of that, the way our brains process these visions is through words; through language and storytelling. Stories about the past and stories about the future allow us to adapt in a complex and shifting world. Stories are how we prepare for the future.

Meanwhile Back at the Bar

"Wow, Dr. Science, that's interesting," James said.

"I'm a futurist, not a scientist." Brian quipped.

"But seriously," James added. "That's *really* interesting."

"I was amazed at how deeply important storytelling was for humans," Brian said. "But, wow, stories are really important to our brains as well. We write our futures before we build them. That's why it's so important to ask ourselves what kind of future we want to live in. I think that steampunk is doing that."

"Yes," James added with a grin. "But steampunk is doing more than that. They're also asking: 'what kind of history do you want to be from?'"

What Kind of History Do You Want to Be From?

James H. Carrott

Imagination and history go hand in hand. Recorded evidence may constitute the bricks of the historical mason's trade, but narrative is the mortar. We're all "do it yourself-ers," building our own histories. History is messy because people are messy. Telling stories about our pasts is a part of being human and how we figure out who we are. This isn't just something we *can* do, it's something we *can't avoid* doing.

Brian will tell you that the future is always in motion, that it's made every day by the actions of people. The past is the same. History is always in motion and it's made and recorded and remembered every day by the actions of people.

There are three fundamental components to understanding history:

1: Gathering evidence

Tracking down references, following paper trails, asking questions. Yes this sometimes means digging through dusty attics and moldy basements in search of stuff no one else has seen in years.

2: Analyzing and synthesizing

Mostly, this is a matter of keeping your eyes open during step one. Patterns will emerge. They may not be what you expected, but everything is part of a pattern. Keep your brain limber; a supple mind can uncover some truly surprising and enlightening things.

3: Storytelling

The heart of history is storytelling. It's not just about searching for other people that came before us. We have to be able to tell it. We have to communicate what we learned.

Technology has dramatically changed each of these three components. But the first has undergone nothing short of an epic change over the last two decades. When I began researching my Ph.D. thesis, there was next to no information related to my topic online. Boy, how the world has changed.

There's just more past now. The past 150 years have changed everything. Cameras allow us to preserve images of people and places. Recorded audio lets us hear others' voices—even long after those people have passed away. Motion pictures bring us another step closer to our predecessors, showing them as living, breathing human beings. Brian says that technology "gives us ghosts from our past;" they live with us and we can talk to them.

The stuff of the 19th century echoes so strongly with us in the 21st, because the 19th century was also the great age of popular printing. New processes made paper cheaper and printing faster; more words bound and shelved each year. As we discussed in Chapter 8, we now find ourselves at a very similar moment. We have more information, we have more history. So what are we going to do with it?

Steampunk looks back into the 19th century and inserts new technologies, stuff that wasn't there before. And that stuff reverberates with the challenges of the Information Revolution and contemporary culture. It hardly seems accidental that our relationship with technology provides an overarching theme for a growing subculture at this historical moment. Immersed in enough of this stuff, a critical mind can't help but turn to questions like "why *this*?" and "why *now*?"

From the outside, steampunk can come across as mere alternative history mixed with a healthy dollop of fantasy. At first glance it looks like quirky romanticized images of goggle-clad airship pirates and clockwork automata sporting top hats. The imagery of steampunk exudes a "rose-colored glasses" sense of turning the dial back, slapping on a layer of the modern technology like robots pouring tea for Victorian ladies and cowboys riding mechanical horses. Even in its darker manifestations, steampunk's quirky aesthetic can suggest a lack of substance. It's fair to say that on the surface, steampunk seems to speak of nostalgic fantasy, changing the past for its own romantic sake. Some of that's true, but there's a lot beneath the surface.

Steampunk is engaged heavily in the third component of my list, storytelling. It recrafts the past, but recrafting the past is no simple project. Play and whimsy can be more than first meets the eye. Those tea-pouring robots and mechanical horses don't just look cool, they raise some pretty interesting "what if?" questions. Changing the past implies changing the present, and more importantly, the future. Even when we rewrite history, without conscious intent, we are also rewriting the future. *With* such intent, new pasts speak volumes about the futures we want to see.

MEANWHILE BACK AT THE BAR

It was getting late. The crowds were thinning out. It was time for bed.

"So what you're telling me is that we need more historians," Brian said, finishing his beer.

"And more futurists," James added, and they both laughed. "I think it's really quite simple and complicated at the same time. People need their technology to be rooted in a past but at the same time they will always play with that past to explore the future."

"We're already seeing it," Brian broke in. "One reporter from the *Wall Street Journal* I was talking to pointed out that Instagram and other photo apps are the perfect mainstream example of what we are talking about."

"Yes!" James yelled. "That's brilliant!"

"People are taking their super high-quality photos and dumbing them down to look old and low res. They use filters called Toy Camera and Retro." Brian searched the bar to get the bill.

"People want their ultra high-res photos to have a sense of history," James grinned. "History is the filter that they are using to process the snapshots of their lives."

"You're funny..." Brian shook his head. "...and yes, you're right."

We Must Design a Better Future

Now that we know people want more from their technology, and now that we know people want their devices to have humor, history, and humanity, what do we do? What do we make? What effect should and will this have on the objects, devices and even buildings we make? Brian has a conversation with legendary designer Bruce Mau and we discover that what steampunk is telling us about technology applies across an even broader design spectrum. Mau and his design team have been working with humor, history, and humanity for some time now and what they created is amazing.

A Conversation with Bruce Mau

Brian David Johnson (Chicago, IL)

EVERYTHING ABOUT Bruce Mau is big. He has a big reputation as one of this generation's leading designers. His collaborators range from legendary architect Frank Geary to big global brands like MTV, the Metropolitan Museum of Art in New York, Coca-Cola, and the Walt Disney Concert Hall. He writes and puts out big books. His 1995 book *S,M,L,XL* is considered one of the most significant design books of the century; with almost 1400 pages, it weighs in at almost 10 pounds. He's even a pretty big guy and commands every stage or podium he stands behind. To top it all off, one of his most famous and impressive museum pieces was called Massive Change, a dizzying collage of print, graphics and installations to put you in the middle of the massive changes that are affecting our planet. His website describes it this way:

> *Design has emerged as one of the world's most powerful forces. It has placed us at the beginning of a new, unprecedented period of human possibility, where all economies and ecologies are becoming global, relational, and interconnected. In order to understand and harness these emerging forces, there is an urgent need to articulate precisely what we are doing to ourselves and to our world. This is the ambition of Massive Change.*
>
> *Massive Change is a celebration of our global capacities but also a cautious look at our limitations. It encompasses the utopian and dystopian possibilities of this emerging world, in which even nature is no longer outside the reach of our manipulation.*
>
> **— BRUCE MAU**
> "Massive Change"

So it's pretty safe to say that Bruce Mau is a big deal. And being such a big deal you would expect him to be a jerk. Being a jerk is kind of an operating hazard for many designers, architects, and even doctors. It's not that hard to understand. If you are going to design and build a multibillion dollar building or cut into another human's body you have to have a pretty strong ego. Let's be clear, you have to believe that you are the one who should be doing it. It's not really a bad thing. Confidence is good. If someone is about to cut into my brain, I'd really rather they have all the confidence in the world. If they also happen to be a jerk, my brain and I think that's just fine... go ahead. Be a jerk, just don't mess up my brain.

But this thing is, Bruce Mau is not a jerk at all. He's surprisingly soft-spoken and deadly earnest.

Before Bruce was a big deal, he was just a young designer from Canada. In the 1990s he started designing the book covers for Zone Press. Throughout that decade Zone books published some of the most interesting and challenging titles like *The Libertine Reader: Eroticism and Enlightenment in Eighteenth-Century France* and *Things That Talk: Object Lessons from Art and Science.* The covers were bold and arresting. During that time I was a ridiculously poor college student in New York City. Standing in the stacks of St. Marks Books (on St. Marks Place between 2nd and 3rd Ave.), I lusted after these books, often skipping meals so that I could save up enough money to buy them.

Just now I went into my library and pulled one off the shelf. I've had it for twenty years; you'd be crazy to get rid of it. It's beautiful! When I flipped to the last page of the book, sure enough there it was: "This edition designed by Bruce Mau."

So to be honest, I was a fan of Bruce Mau before I even knew who Bruce Mau was.

In 1995 Bruce teamed up with renowned Dutch architect and urbanist Rem Koolhaus to produce a 1376-page-long collection of essays, diary excerpts, travelogues, photographs, architectural plans, sketches, and cartoons. The book, *S,M,L,XL* was a huge success, selling out multiple editions. It was even counterfeited in China and Iran. (Now that's a good book!)

But at the beginning of the 21st century everything changed. The Vancouver Art Gallery commissioned Bruce and his team to bring out the show Massive Change. The exhibit was on display at the Vancouver Art Gallery for three months from October 2, 2004 to January 3, 2005. From there, the exhibit went to the Art Gallery of Ontario in Toronto for three months from March 11 to May 29, 2005. The exhibit was on display at the Museum of Contemporary Art in Chicago from September 16, 2005 to December 31, 2006.

Since that time Bruce and his design team at Bruce Mau Design have been crusaders for massive change. They have launched education projects, socially conscious design workshops and a global commitment. Throughout his crazy schedule Bruce adheres to a single rule: We must design a better future.

"Ok Bruce, how do we design a better future?" I started.

"It's a real challenge," Bruce began. "When most people think about design they imagine hefty, expensive objects created by singular authors, who are responsible for the whole thing. They think about designers picking colors and shapes. But that's not really what designers should do. That's not how I think about it. It's not how I design.

"When I talk about design I think of it as *big D design*." Bruce paused, took a breath, and continued, "Real design innovation has nothing to do with what it looks like but what it makes possible. What's the new capacity that we're developing? Design is leadership, the capacity to imagine a future and then systematically articulate the future. This is what leaders do. They help us imagine a future and they work to systematically execute that future and build it.

"I started as a communication designer in Canada. If you had told me 25 years ago that I was going to be going what I'm doing now I would have been extremely skeptical. I couldn't have imagined that this is what design is all about. But amazing things have happened in my lifetime and in my career. Communication went from being something that happens after all the decisions were made to people realizing that communication is actually strategy. We started to understand the effect that the design was having on people's lives and we started design for that. We stopped designing the objects and we started design the effect."

The Incomplete Manifesto for Growth

In 1998 Bruce wrote a list to articulate his beliefs, strategies, and motivations behind design and life. He still uses them today as the guiding principles and design process for himself and his design studio.

Design Is Humor

"I want to talk to you about humor," I started. "As we explored steampunk culture, we learned that people wanted to have a very different relationship with their technology. People wanted their technology and devices to have a sense of humor. How do you use humor in your work? What would it mean to have technology with a sense of humor?"

"There are two ways that humor can be a part of design," Bruce replied. "First, in our culture humor is the place of truth. It's not an accident that the most respected sources of news today are *The Daily Show* with Jon Stewart and *The Colbert Report*."

"The fact that these shows are so popular tells you something," Bruce explained. "In our culture today people have a higher regard for the news coming from comics than they do coming from the traditional sources. Jon Stewart was once criticized that he wasn't always fair and balanced on his show. He told the interviewer from The Fox News Network that he's a comic—his show is on the Comedy Central cable network and often times his show came on after a show that makes prank phone calls. Stewart went further and said it's an indictment of our traditional news system that people think he's more trustworthy as a source of information than a real news network.

"The power of humor is the reason that Jon Stewart is able to tell the truth. Both Stewart and Colbert can use humor to address important issues that are too hard to address directly. They socialize public discourse in a way that is friendly and open and not so brutal. But really if you think about it, it's more brutal than a classical debate.

"The second way that humor can be used in design is as a way of thinking," Bruce continued. "I love anything to do with puns. For me humor and jokes demonstrate a higher order of thinking. When you can maintain your sense of humor then you can maintain a higher level of discourse and discussion. You can not only address the issue, but you can actually deal with it.

"From a design perspective, humor allows us to see things in a way that we wouldn't have been able to do before. You can flip it around, look at the problem from different angles and have fun with it. In my design shop we use humor all the time. It's an essential way of thinking. Design *is* humor."

"Do you think that every time you design something, whether it be a book or a library, is it always with a sense of humor?" I asked.

"I always try to work with a sense of humor," Brice replied quickly. "But I don't think every design needs to be so humorous that it makes you laugh. It can also produce delight. Delight comes from the ability to make things, inventing things, and moving things around in a way that demonstrates intelligence. That definition of delight and humor is very close to what I try to do in everything that I design.

"I became conscious of this in my own studio. If you read the *Incomplete Manifesto for Growth*, one of the points is *Laugh*. Over the years I started to notice that people were always commenting that we laughed a lot in my studio. I also noticed that there were times when we weren't laughing. During the times when we were laughing, we were doing our best work. We were at our best. The times when we weren't laughing was an indicator that something was wrong."

"As a designer have you seen examples of good designs that are based on humor?" I asked.

"The Colbert Report is a great example. Colbert is an invention of Stephen Colbert. The character is such an extraordinary piece of design. He is so carefully created. Colbert made himself an invention and then that invention produces all kinds of different ideas and stories. Colbert produces all the other daily events on the show but he himself is a project. That kind of design is brilliant.

"I have seen some clever work in the product realm," Bruce shifted gear. "I work with a company called Emeco. They wanted to make a classic mid-century designed chair but wanted it to be from recycled materials. We teamed up with Coca-Cola to make the 111 Navy chair. It's a beautiful chair that looks like a work of art but at the same time it's made for 111 recycled plastic bottles. Every chair that they produce takes 111 pop bottles out of environment. That's a higher order of thinking.

"But what's clever," he added, "is that they made it look like the classic chair. It's delightful to sit in. It's a great chair. You don't need to know that it's made from 111 recycled bottles but when you do... then you smile. It's something extra; it's clever and you do get delight from it because you're doing something good."

History Is a Bridge to New Places

Bruce's take on history is different than most people's. He sees the pragmatic effect that history has on the things we design today. Yesterday is the foundation for today so that we can build a better tomorrow... only sometimes you need to reach a little further back into history to find a solid and positive foundation.

"What role has history played in your design?" I asked.

Bruce answered my question about history with a reply about... science! "We have reached a point where science and our knowledge of the universe has separated from our personal experience. Up until Newtonian physics you could experience and understand the universe. It was physical. It had a kind of mechanical correlation."

Newtonian Physics

In 1685 when Isaac Newton was 23 his school (Cambridge University) was shut down to protect against the Great Plague. Isaac was not a great student at Cambridge, but working on his own at his family home in the country, he would change the world of math, physics, and all of science in general. The work he began when he was away from school he would later publish in the book called *Mathematical Principles of Natural Philosophy* or the *Principia* for short. In it he outlined his Three Laws of Motion. Here they are—first in technical language, then I had a little fun explaining them.

1: The velocity of a body remains constant unless the body is acted upon by an external force.

Or simply: A rolling ball will keep rolling until someone messes with it.

2: The acceleration of a body is parallel and directly proportional to the net force F and inversely proportional to the mass m, i.e., F = ma.

Or simply: A rolling ball will change course if two people are pulling on it, and it will move in the direction of the person who is pulling harder.

3: The mutual forces of action and reaction between two bodies are equal, opposite, and collinear.

Or simply: If you and I both are pulling on the ball, then you and I will feel each other fighting over that stupid ball.

Now all this talk about the balls was a big deal. Of course Isaac didn't write like that—he actually wrote the entire *Principia* in Latin so only "serious" people would read it. But all this ball talk really did change the world. People for centuries used the three laws to understand things like motion, gravity, and how the planets moved in the sky. Isaac's laws explained what we were seeing when we looked at balls rolling around on the floor or when we watched the planets in the night sky.

"We've come quite a long way since that time," Bruce said. "In my lifetime we have lost that mechanical correlation. Our understanding of the universe has completely changed. We can't see it anymore and in many cases it contradicts Newton. Quantum Physics turned Newton upside down. It's harder and harder for people to understand the world we live in. You can apply this beyond just physics. It's true for most people and it applies to technology, politics, and so many parts of our lives."

Quantum Physics

Quantum physics is spooky. No, it really is. Albert Einstein, Mr. E=MC2, described what he observed when he first saw the effects of quantum physics as "spooky." Why is it spooky? Well, first it doesn't adhere to Newton's three laws. It does crazy stuff. It deals on a nearly invisible scale that is atomic and subatomic.

British physicist David Deutsch is a pioneer in the fields of quantum computing and the theory of multiple universes. Deutsch explains quantum theory like this in his 2011 book:

> *"Quantum theory is the deepest explanation known to science. It violates many of the assumption of common sense, and of all previous science—including some that no one suspected were being made at all until quantum theory came along and contradicted them. And yet the seemingly alien territory is the reality of which we and everything we experience are part. There is no other."*
>
> **— DAVID DEUTSCH**
>
> *The Beginning of Infinity* (2011)

"The fact that we can't see it is a big deal. We are so distant from our understanding of the world around us that it produces a need for stability. People have to find a way of to be okay in this new world." Bruce's voice filled with passion. "This is a significant problem that we have to respect. We need to understand it so we can design for it."

"This is what got me thinking about designing social change. If you begin to design social change, you realize that the object of the design is no longer just a product. We are designing an experience that can reinforce people's feeling of stability. We can make them feel better and allow them to embrace innovation and change. If you want to innovate you need to give people anchors. You build and innovate from there."

"I can give you an example," Bruce explained. "When the Internet came along it was new and we didn't have a way to talk about it so we borrowed previous language and metaphors. We used the language of the book as our anchor. We talked about pages and folders and trash cans. This allowed millions of people to take on a new way of thinking, a new way of seeing the world around them. This was profoundly important. We need to explore this new space of technology. As a designer I use history as a bridge to new places."

Extraordinary Human Capacity

Deep in Bruce's heart, he is a raging optimist. He believes vehemently in the power of design to change the world. When we started talking about how people wanted their devices to have a greater sense of humanity this touched a chord in him. We are at an interesting point in history where we as humans have access to technology and many realize that we can fundamentally affect the future. Designers are actively pondering the possible futures that they want to live in and then using all of this technology to build that future. This is new. I asked Bruce what he thought about what was happening today.

"I love the optimism in your perspective!" Bruce yelled. "I'm deeply optimistic in my work. As designers we can't afford the luxury of cynicism. Thinking about the future is a fundamentally optimistic act. It's a powerful thing to think about the future as something that is *inventible.* That's what design is all about. It's also the DNA of leadership. If you have the ability to imagine a future and change it, then you can be a leader.

"Something that troubles me recently is that I think we are producing two classes of people: a consumer class and a producer class," Bruce explained. "The consumer class is very different from the producer class, one makes and one consumes. One has the power to build while the other is powerless to act. What worries me about the producer class is that they typically interact with other producers. It is our responsibility to prevent that class structure from crystallizing.

"I come from a mining town in northern Canada," he continued. "I never heard the word *design* until I left that town. For a lot of people where I grew up, they interact with the world through consumption. They don't produce things. Wait, that's not 100% true—they produce all sorts of things locally. They're hunters, they build their own homes, they build their farms, they do all kinds physical and local production. But when it comes to a modern global life they are consumers. They wouldn't even think about being a producer. They buy a phone and they talk on the phone. They never imagine actually making a phone. They don't even have the skills to imagine what it would be like to make a phone.

"My mother was very much in the consumer class. She was largely subject to the world and to the producer class who had all power. She never had access to the resources and knowledge to do anything different.

"But I am an optimist," Bruce said. "This is the best time in human history to be alive. Humans have a capacity to do incredible things. We have this extraordinary capacity to make change and build things. One problem is that so many people express themselves through consumption. We need to change that. They don't have the tools and the skills to do anything different. They aren't in control of their own future. This is why we are working on Massive Change. That's why we are building the Massive Change Network. The real promise of the technology is its distributed power. We need to reach the consumers who've never had access and unfortunately many of them are fearful because they have never been confronted with these ideas. It's a different ball game and it's a really challenging issue."

"So is it a mindset shift?" I asked. "You want to make sure that people don't just sit back and let the future happen to them?"

"That's a powerful idea." Bruce paused and sat with the idea for a second. "That's a huge idea! If there was one thing we could do, if there was one change we could make in the world, we should provide people with agency. Provide them with the tools to imagine a different future for their own community.

"I have to tell you about a project we're doing in Guatemala. This project completely blows the doors off of any definition of what design can accomplish. With this kind of thinking about history and the future we can make a big change."

Bruce settled down and started again, "A group of people in Guatemala contacted me and said: Our citizens have suffered through 36 years of civil war. When our citizens think about the future, the vision that they have is dominated by three images: violence, poverty, and corruption. So when they try to imagine what their lives will be like in 10 years what they imagine is more violence, poverty, and corruption. We need to change that. We need to work on inventing an image of the future that's more powerful in a positive sense. We need the positive to dominate the negative. We have to work towards this new reality. We need to produce a vision of Guatemala that will allow people to dream."

"That's amazing," I said.

"It's a long-term project that's still going on," Bruce continued with pride. "It's called *Guapeamala*; the love of Guape, the love of this place. We've developed what we call the culture of life. After 36 years of the culture of death, we wanted to now build a culture of life. It's not like you can just turn that kind of thing on. You have to actually design it.

"I think we can apply the same idea to America. We have built foundations in our culture that we take for granted. We take for granted a culture of justice, a culture of education, a culture of entrepreneurship, and a culture of dreaming. But we take all of that for granted. But some of us have taken the responsibility for dreaming, that's our job, to dream. We contribute to society by dreaming. We dream about positive things, optimistic futures, and then we work to realize those dreams.

"After 36 years of atrocities, Guatemala had lost that ability to dream. To fix that we needed to build a solid foundation. This goes to your idea that people want a sense of history. The Guatemalan people needed a history that they can look back to, past that terrible 36 year period. As a part of this project we had to search out and foster a foundation that was positive. We wanted people to have the capacity to reach back and seek human society in a positive way. That's what history can do; history can empower an entire country of people. That's incredibly powerful."

"The way that you change the future..." I started. "The way that you change the future is that you change the story that people tell themselves about the future that they are going to live in. On one hand that's really easy but on the other it's incredibly large."

"That's exactly what we're doing with Massive Action!" Bruce replied. "Everything we do follows that ethic. How do we put the tools in the hands of as many people as possible as quickly as possible for the lowest possible cost?

"There are so many forces that deny agency to most people. They try to concentrate that agency in the smallest number of hands possible. Just look at our universities here in the United States. Our universities advertise how exclusive they are. It's horrendous! They measure it and they use it as a form of advertising. They use it to promote the university and for some crazy reason the more exclusive a university is the better. The more young kids they turn down, the better and more valuable they think the place is. That's insane. If we are going to make real change the universities have to be on our side.

"We need to understand what's our real purpose," Bruce finished. "What we have in front of us is big. The opportunity is massive. This period in history will determine the coming millennia."

The Incomplete Manifesto for Growth

43: Power to the people

Play can only happen when people feel they have control over their lives. We can't be free agents if we're not free.

Humanity in the Machine

James tells us how he first met Brian and how they bonded over technology, humanity and... hair! Then our historian relates a particularly memorable conversation with Shanna Germainis when she asked, "Where does the Internet live when it goes home for the night?" Then Brian and Byrd fly down to Silicon Valley to talk with Bruce Sterling—not only one of the founders of steampunk, but one of this generation's leading thinkers about the future.

As we close James tells us about "Mr. Jobs's Incredible Music Shrinking Apparatus" and how we just might be able to build a future we want to live in, a future that can stand the test of time.

Humans (and Steampunks) are Hairy

James H. Carrott (Seattle, WA)

WE HUMANS NEED to bend our technologies and stories—our pasts, presents, as well as our futures—so that they relate to us. We need to give ourselves room to be people in a world brimming with technology. In the end we're the ones who build the future. We build it out of the stuff of imagination.

I think that in the end, it comes back to hair. Humans are hairy. We leave bits of ourselves behind wherever we go. Just check your vacuum cleaner bag or your shower drain if you have any doubts. Our hair is one of our most distinguishing qualities—color, texture, length, bodily location. That's not even getting started with what we do to and with it—shaving, styling, coloring, cutting, tweaking, and tweezing. No matter how hard we try to hide it, hair is one of the things that reminds us that we really are apes after all. We're all different, quirky, and unique and there's never really one size that fits all.

When we sat down for the first time to begin the process of turning our research into a book, Brian told me a story about computers and hair that's become one of my favorite anecdotes. As I recall, he started off like this:

"One of the problems that I have as a futurist is that people typically imagine the future as hairless."

Coming from Brian, this statement in itself is hysterical. His completely bald head, clean-shaven face, and perennially boyish grin are almost trademarks. He's a kind of ambassador for the future; always asking "Well, what future do you *want*?"

"There's a problem with the idea of a future without hair, though," Brian took a sip of beer and grinned mischievously. He's got a knack for that. "And there's a story about it."

There usually is.

"I have a very good friend. His name is John Croft and he's a software engineer," Brian continued. "We build computers together. Many years ago he and I built a desktop machine and I asked him 'should I put it on the floor or should I put it on the desk?' John had no idea what I was talking about. He shot me this look like he thought I was crazy; like only a lunatic would ask a question like that.

"So I said, 'I'm asking if should I set it on the floor or on the desk, because putting it on the floor gives me more space, but it seems like I should sit it on the desk because it would get dust and things like that if it was on the floor.' John looked at me again with a scoff, laughed in my face, and asked, 'Have you ever opened a computer that's been in a house for more than a few years?' I hadn't, so I answered, 'No.' He said, 'Don't do it. They're disgusting.'

"He elaborated. It turns out that after a few years in a house full of human beings, even a sleek and beautiful computer gets all full of dust and pet hair and skin and all of this really disgusting stuff. Houses are hairy. They're where people live. And I thought: 'That's awesome!' I loved that idea that houses are hairy. It has strongly influenced the way I think about the future. Humans are hairy. Our lives are hairy; they're full of mess and junk, quirky twitches and broken bits. Our lives, like our houses, are not streamlined and perfect or buffed to a high sheen.

"A look inside the computer tells us that the stuff..." Brian finished up. "...all the stuff that makes us human, is hairy. Humans are complicated, we bring a lot of stuff along with us. Usually when people envision the future, it's a flawed vision, because we don't envision a future with hair and dust and mess. We envision a future that is cleaned of all that, and thus raw and intellectually bankrupt and false. It's not human. Humans are hairy."

It's this kind of thinking that led me to Brian in the first place. I heard him speak at the Consumer Electronics Show in Las Vegas. It's a huge industry show —and when I say that, I really mean huge. It's the convention of conventions for consumer electronics. Everyone and their mother company goes there to show off their latest and greatest. It's a place of gadgets, shiny lights, "Version 3.0! Now even newer!" It's the kind of place where, when you attend a talk on "The Future of Television" you expect corporate-speak and technobabble—lots of showy tricks and trite marketing gimmicks.

What you don't expect is Brian. He talks about people. He frames ideas clearly and explains his thinking. And there's always a story, peppered with "...and that's awesome!" and such childlike grins that you can't help but be drawn along into his sense of wonder. He understands, on some eerily intuitive level, that even when he's talking about the most complicated technology, he's talking to people about people. His talk really moved me. I remember thinking: "Wow, that *really is* awesome. He gets it. I need to talk to this guy."

I went up to the podium once his talk was done, gave him my card and asked if he wanted to grab a beer sometime. The rest, as they say, is history.

Remember! A Gentleman Grows a Beard! from Fable 3. (image courtesy of Lionhead Studios)

Great Hairy Steampunks!

Steampunk worlds are openly false, but they embrace hair. It's hard not to embrace hair when the 19th century, a period thought of by many as "the great age of facial hair." Steampunk is built on fantasy and imagination, but it's not clean-shaven and sleek. Steampunk plays with its hair; waxing great big moustaches into curly-queue points, fluffing monstrous mutton chops into facial forests, teasing long braids up into sculptures so epic they need only the tiniest of hats, bobby-pinned at a jaunty angle.

I shaved off my own beard partway through the summer of 2011. I needed to make a change, and that was an easy one to do. I grew it back like I always do, but it still amuses me that I appear both bearded and clean-shaven in the Byrd's *Vintage Tomorrows* documentary. At a sneak-preview screenings that October, an audience member asked director Byrd McDonald and I what we had learned in making the film. I jokingly responded that I'd learned that I shouldn't shave. That drew a laugh (really, I'm 40 and look almost like a kid without my beard), but when I thought on it more, it occurred to me that our hair is a part of who we are—and everybody's hair is different. Speaking generally of what's socially acceptable in mainstream culture, women tend to have more freedom around how they wear the hair on top of their head. Much of men's freedom lies in what we do or don't grow on our faces.

So, like I do when I'm interested in something, I did a little research. It turns out that up until the 20th century, who was "permitted" to grow facial hair and how they were supposed to grow it was largely directed by class, religion, occupation, or other external community standards. The term "barbarian" originally translated to "bearded ones"—shaving was a way the Romans distinguished themselves from savages. On the opposite end of the spectrum, Victorian gentlemen viewed robust facial hair as a sign of healthy masculinity, and splendiferous whiskers became the signature of many prominent scholars, politicians, and warriors. During the US Civil War, Union General Ambrose Burnsides lent his name to his own branded trim—"sideburns."

Senator Ambrose Burnside of Rhode Island (photo courtesy of Matthew Brady and Levin Corbin Handy, (ca 1880))

Josh Tanenbaum as Doctor Watson. (photo courtesy of Karen Tanenbaum)

But the grand age of bushy faces came to an end at the turn of the century with the advent of the Gillette disposable safety blade. Troops in both World Wars were issued razors, and our noble boys sent off to war looked more like the boys they were than in the previous century. After World War II, goatees became a bohemian standard (ask people to describe a 'beatnik" to this day, and they'll talk of goatees, berets, black turtlenecks, and bongo drums), and the long, full beard grew back into prominence as a symbol of the "hippie" rebellion. Both countercultures chose a scruffy rebuff to *The Man in the Grey Flannel Suit*—their facial hair distinguished them as individuals in a world of clean-shaven suburban clock-punchers.

Once razor blade marketers in the 1920s convinced women that underarm hair was unpleasant and unfeminine, and fashion trends shifted toward sleeveless or sheer-sleeved dresses, western women had entered the game in earnest as well. Leg shaving never gained the same momentum, but higher hemlines and sleek-legged icons like Betty Grable made the practice a lot more popular in the US and Britain by mid-century. The practice of women shaving body hair became so prevalent in the Anglo world that not shaving one's legs or underarms became a kind of symbol of feminist rebellion toward the end of the 20th century.

You may think I've wandered off the point, but I haven't. At least not so far that I can't bring us back. You see, growing hair is something all humans do. Even after death, for a while at least. No two humans grow hair in exactly the same way, and it's something that visually distinguishes us as individuals. We leave hair everywhere we go. It is, in its own way, a kind of history. Whether it's caught up by cooling fans and sucked into the innards of our computers and gaming systems, washed down the drain, snagged in a brush, or clipped off in the barber's chair (a nod again to Lancer here), hair defines us. And steampunks are kind of obsessed with it. Male steampunks sport the widest, cleverest array of whiskers our society has seen in a hundred years.

Steampunk's emphasis on individuality manifests itself in flamboyant costumes, handcrafted oddities, and wondrous machines, all made from the cast-away stuff of contemporary consumer society. Culturally speaking, steampunks take the swept-up clippings off the salon floor and make them into things of beauty—spinning the waste of conformity into threads of doggedly individualist craftsmanship.

The stuff of Steampunk is the stuff of hair, the stuff of being human.It puts an emphasis on the value of humanity, an individual humanity in an age of mechanization, development, progress, mass production, mass consumption, and dizzying technological innovation. Steampunk embraces things and people simply because they are quirky and interesting, odd and different.

Where Does the Internet Live When it Goes Home for the Night?

James H. Carrott (Portland, OR)

Shanna Germain is best known as an author of erotica. She writes smart, well-crafted erotica, which does in fact exist, despite our mainstream culture's usual spin on the topic. This makes her an expert in some fundamentally human stuff: love and lust, hearts and loins, passion and play. She also thinks a lot about technology and its role in our lives (and in our bodies). I've spoken to a lot of smart and provocative people in my travels, but this conversation remains one of the most enlightening. Both Shanna and her insights are beautiful without pretense, thoughtful without "Theory," and deeply human, with all the complexity and character that entails.

We met for coffee in Portland, Oregon on a Monday morning after the Gearcon convention. I had a train to catch back to Seattle and she had a manuscript to edit, but we carved out a little time to chat. We took our coffees into the sun, switched off our cell-phones, and talked. That half-hour sitting at a rickety iron mesh table in an echoey courtyard brought me back to square one: What does it mean to be human? What are we going to do with all this technology? How do we create the human future we want and need?

"One of the things I'm interested in is bringing the fringes to the mainstream," she started stirring her coffee. "It's still the same emotional truths, no matter your class, sexuality, race, or even who you are as an individual. At one level we all want the same things: We want to be liked, we want to be loved, we want to feel comfortable and secure.

"I think steampunk is going to keep growing in the mainstream," she continued. "We'll keep the heart of what is real and true to us as humans and then expand outwards to see how far we can push it. We can humanize technology; give faces, names, and personalities to mechanical devices. Even today most people name their laptops. We need that connection."

I thought of Molly Michelle Friedrich's "Mechanical Womb with Clockwork Fetus," which renders in technology one of the most human processes imaginable: reproduction. A deeper impulse seems to run through the intersection of humanity and technology. I asked: "When you talk about creating a human face for technology, naming a laptop, do you think that's a part of what steampunk overall is trying to do?"

"I do," she replied bluntly. "There is something scary about not being able to understand how things work. Especially not being able to understand how things work when we use them all the time. When I first heard about the internet, *the cloud*". She emphasized the word. "I thought that's really interesting. Someone had to name it *the cloud* so that people could have a visualization of this thing that has no visualization. A few years ago I was trying to get someone to explain to me: What is the Internet? What is it? What is it built out of and where does it stay? What's the pet place where it lives?"

"Where does it go when it goes home at night?" I smiled.

"Yeah, totally." Shanna smiled back. "Who named it? Since then I've been interested in the number of ways people are attempting to describe the Internet.

Molly Michelle Friedrich, Mechanical Womb with Clockwork Fetus (2009)

Everyone seems to try and talk about it as a tangible thing that we know. Some people describe it as a solar system and others say 'imagine a village with all these houses...' The technology we have all around us is something we don't even have the words to encapsulate or imagine."

That idea has stuck with me ever since. Steampunk helps us construct the images we need to understand our technology. It makes technology more human so that we relate to it, and engage with other humans about it. Plus, who wouldn't want to know where the Internet lives when it goes home for the night?

Fantastic, Marvelous, and Wonderful Technology Being Used by a Bored 16 Year-old Girl

Brian David Johnson (Silicon Valley, CA)

I was on the early flight into San Jose, Silicon Valley's airport. Up until a few years ago it was charmingly small and cramped—and by charming I do mean it was really hard to fly in and out of. It always smelled of some guy's old pepperoni pizza and stale coffee. Happily, the new terminal opened up recently and San Jose now feels like every other modern airport. It's clean, efficient, and has a Ruby Tuesday.

Jumping in a cab I headed to the San Jose convention center where I was going to meet up with Bruce Sterling. Besides being one of the authors (with William Gibson) of one of the most important steampunk books, *The Difference Engine*, Bruce was also one of the first great popularizers of the cyberpunk science fiction movement of the 1980s. For years Bruce would tell interviewers who asked him about his futuristic visions becoming a reality that he was simply a science fiction author. He wasn't a futurist or a designer. He was a writer whose job it was to entertain the reader. That worked for a while, then something happened. Bruce started thinking seriously about the future.

In 2002, Bruce published *Tomorrow Now: Envisioning the next 50 years*. In it Bruce explores possible futures as well as changes in culture, business, and politics. He set himself a clear goal for success. In the introduction he writes that true success for a futurist is when their ideas become clichéd. When they are so common that people just assume that they have always been there. "Nothing obsolesces like *the future*", he wrote, "Nothing burns out quite so quickly as a high-tech avant-garde. Technology doesn't glide into the ranged, streamlined world of tomorrow. It jolts and limps, all crutches and stilts..."

Since the publication of *Tomorrow Now* Bruce has been active in the field of design. I met him that day at the Augmented Reality Event. Bruce is one of the guiding visionaries for work being done in this area. Augmented reality blends the physical world with the digital world, usually through computer screens or glasses. The folks putting on the event were gracious enough to let us in the day before the event so Bruce and I could sit down and chat and Byrd and the documentary crew could film it.

During that afternoon we talked about steampunk and *The Difference Engine*; all things that you've read previously in the book. But at the end of our time together we started talking about the future.

Bruce Sterling and Brian David Johnson chat at the Augmented Reality Event in Silicon Valley, California. (photo courtesy of Byrd McDonald)

"As a part of my research as a futurist," I began, "I always ask people what kind of future they want to live in and also what kind of future they want to avoid."

"Yeah," Bruce chimed in with his unmistakable Austin, Texas accent.

"You've been writing and thinking and designing the future for a while now," I continued. "What kind of future do you want to live in?"

"If you had asked me this back in the '80s, I would have told you that I wanted a future where everyone had a personal computer," Bruce laughed. "And well... that happened. Actually it more than happened. I've had several personal computers and now we are in a time when the computers, the devices don't matter anymore. It's about the software and the experience." Bruce rubbed his chin and thought for a second. "It's hard to have envisioned a future and then gotten it and then moved on."

The sun was starting to go down and Bruce had flown in that morning from Italy so I figured we didn't have much time left. "So what kind of future do you want?"

"I really think that I'm going to live to see a time in the next decade when we will stop surprising ourselves every 18 months with a new gadget. People live longer than 18 months and we need to have a wiser, more long-term vision. Sense of wonder has a short shelf life.

"I want to see a future where our relationship with technology is more humanely metabolized. It doesn't have to be an awesome wonder because it's just a part of the fabric of our lives. Awesome wonder should be reserved for really important things like the size of the universe, not our devices and technology."

"Do you think it's more of the human content?" I asked. "With our research in *Vintage Tomorrows* we've seen that people want their technology to have a sense of humanity."

"Yeah! When I write about technology," Bruce explained, "I want to drag the reader into a more intimate relationship with that technology. I want it to live in their sensorium, I want it to live next door. One of the best things you can do if you have something fantastic, marvelous, and wonderful is to have it used in the story by a bored 16-year-old girl. It engages people in the future in a more aware fashion. It's more honest, more modern."

We finished up the interview and Byrd gave me the thumbs up that he had what he needed. The Augmented Reality crew showed up to take Bruce off to dinner and I needed to head back to the airport.

What struck me most about Bruce's vision for the future was that it was so firmly based in humanity, in the everyday—even in the hands of a bored teenager. It echoed what we had learned all along our journey through steampunk: People want their technology to have a sense of humanity.

At the end of Bruce's Introduction to *Tomorrow Now*, he wrote a great last line. Here you have Bruce Sterling, one of the leading thinkers about steampunk, cyberpunk, augmented reality, and our technological future, and how does he choose to end his first big work on the future of technology? He writes simply, "The future will be lived."

Mr. Jobs's Incredible Music-Shrinking Apparatus

James H. Carrott (Seattle, WA)

Steampunk shows us that humanity needs to be at the center of our future, at the heart of the technologies we build. That's not just lip-service to the importance of thinking about people, nor is it a statement about usability or interaction. It's something fundamental to the nature of the future we build: it needs to be human.

The devices we create need to have a meaningful element of humanity in them. They need to be clearly built both for and by people. Steampunk underlines the character in objects and highlights the fact that the things with which we surround ourselves have character. Whether we like it or not, we have relationships with objects. Steampunk insists that those relationships can be warm, deep and meaningful; not cold, distant, and disposable.

Greg Broadmore's hilarious and masterful book *Doctor Grordbort's Contrapulatronic Dingus Directory* is one of the purest examples of this.

Greg's directory of fantabulous devices especially shows how steampunk devices, art, and technology are all named. They come from someone. Someone (in this case Doctor Grordbort) made them and gave them insanely funny and descriptive names. Steampunk wants a world where our technology is not only designed for us but it's also *made by* us. There is humanity in the device; wit, humor, and ingenuity that brings about amazing things.

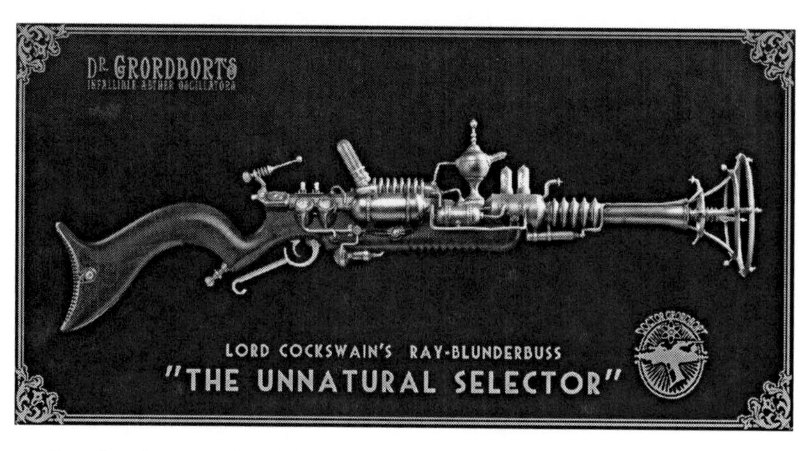

Lord Cockswain's Ray-Blunderbuss "The Unnatural Selector" © Stardog and Greg Broadmore

Just look at all the great devices we have today in the 21st Century. Many of the most popular and loved devices are tied very specifically to people. How many of us have not marveled at Mr. Jobs's Incredible Music-Shrinking Apparatus? Your entire music collection on one tiny device!

Steampunk physically demonstrates that if the future we build is completely clean, sterile, and streamlined, it loses a lot of its meaning. When we envision our future we need to remember who we are: quirky, hairy, flawed human beings. If the things, processes, even social structures we create miss that, they miss us. If we put people at the center, the tools we create can inspire and empower. If not, they can just as easily alienate and oppress. Steampunk suggests that we can build a future with human character, one that recognizes the strengths of our differences. If we do that, then we're building a future we can inhabit—a future that just might stand the test of time.

We Want to Remember a Time When Our Lives Were Not Made of Plastic

Vintage Tomorrows was pretty much done. Byrd and the documentary crew were editing away. Most of the book was written and this was the boys' last interview. James is driving south to Burning Man and he stops off in Portland to see Brian. Hunkered down to work at a hotel, the two take the opportunity to cap off the book with a phone call to Jake von Slatt.

Jake is a maker extraordinaire. Through the entire Vintage Tomorrows journey, whenever anyone talked about making steampunk things, Jake von Slatt,proprietor of the Steampunk Workshop, was the first name out of their mouths. Bonus for the boys, Jake gave them the artwork for the front and back of this book! Jake's magnificent creations perfectly capture the spirit of the book. The past and the future. Hand-crafted objects and technology. All wrapped up in gorgeous steampunk splendor.

James and Brian bring their journey full circle talking to the maker's maker about what they learned. What does Jake think about humor, history, and humanity?

On the Road to Burning Man

James H. Carrott and Brian David Johnson (Portland, OR)

"Made it to town. Parking car now. I'm in room 512," James texted Brian.

"SWEET! On my way," Brian replied back.

Fifteen minutes later Brian rang the doorbell of room 512.

James answered with a bewildered look at the door. "I have a doorbell."

"Yes," Brian answered. "I just rang it."

"I have a door bell," James said again, now sounding like a crazy person. "I didn't know I had a doorbell. Come in," he ushered Brian into the room. "You'll have to excuse me." James pushed his glasses back on his nose. "I'm running on three hours sleep. I was packing for the Burn. I have enough gear in my car to live through two apocalypses."

James was headed south to the Nevada desert for Burning Man, the yearly festival of freaks, geeks, and fabulous contraptions like the Neverwas Haul that started James thinking about steampunk and technology and set them off on their journey. It was actually kind of perfect to do the final interview this way. With James headed back to Burning Man and Brian flying out the next morning to give a keynote address about the future of technology.

"You have to see my new goggles," James said proudly. He gingerly lifted a set of thick goggles from a tin case. Removing his trademark circular cultural historian glasses he popped the goggles on his head. "Check these out. I needed some heavy duty goggles for the burn. The dust out there is so fine it gets into everything. Kimric (we met him in Chapter 11) made these for me."

Brian smiled, "Wow they are something." James looked a little crazy. "Wait, are the lenses in those prescription?"

James blinked back at him and smiled, his eyeballs magnified.

"That's awesome!" Brian shouted.

"I know," James slipped the goggles off and replaced his glasses. "Should we call Jake before I fall asleep?"

James's new goggles. Kimric Smythe designed and built them to withstand Playa Dust while accounting for an astigmatism and bifocal lenses (early American history has its risks—years of reading 300-year-old pencil and quill manuscripts can do a number on your eyes). They worked like a charm. (photo courtesy of Steve Brown)

The Maker's Maker: Jake von Slatt

In July 2007, Jake von Slatt and his "Wimshurst Machine" adorned the cover of *MAKE* magazine. The article was titled, "Meet Mr. Steampunk: Jake von Slatt." For many people Jake is the steampunk maker's maker, and although he's a humble guy and does a great deal more than steampunk, he's become the poster boy for steampunk makers all over the world. We wanted to talk with Jake about his beautiful devices and to ask him what he thought about what we learned.

"Hello, Jake." Brian leaned over the mobile phone and said into the speaker, "This is our last interview for the book. James is blowing through town on his way to Burning Man with enough camping gear to make it through two apoc..."

"Aww, I can't go this year," Jake broke in.

"I'll be with the Neverwas Haul crew too." James smiled. He was obviously excited for his trip.

"Ohhh," Jake groaned. "It wasn't there last year. It was in dry dock getting repaired."

"This year is the grand re-launch!" James added.

"The one year I'm not there everybody's going to Burning Man except me!" Jake moaned again.

"Well, we'll both be there next year," James reassured him.

"Okay fellas," Brian broke up the Burning Man frenzy. "Jake, we wanted you to be the last interview for this funny little project. It's become an M.C. Escher kind of thing. James and I started off by asking, 'What can steampunk teach us about the future of technology?' Then midway through the project Byrd McDonald, a filmmaker, decided to make a documentary based on the book, but he also wanted to document us finishing the book. So now we have a documentary based upon the book that's also about us finishing the book..." Brian smiled at James as Jake laughed on the phone.

"Wow," Jake said.

"I know." Brian continued, "So James and I decided, well the heck with it, we are going to write the documentary into the book. So now Byrd and the documentary crew are in the book, and *Vintage Tomorrows* has become this crazy self-reflexive thing where you don't know where one part stops and the other begins. You know, like two hands drawing each other." Brian acted out the famed Escher print with his hands.

"It sounds like it," Jake chuckled.

"So Jake, we thought it fitting to interview you," James added. "You gave us the perfect iconic image for this book. We wanted to chat about that and also about what we've learned making this book." James took a deep drink of coffee. The three hours of sleep were beginning to show on his face.

"Let's start with your wonderful device," Brian started. "Jake von Slatt, what is this beautiful apparatus you have given us for our book?"

"It's an all-in-one PC," Jake started casually. "For a computer, it's a pretty simple hardware setup. That's a 24-inch monitor with a full-size motherboard, hard drives, and video card attached to the back. But the inspiration was the theater."

"The theater?" James asked.

"Yeah, I wanted it to look like a theater stage," Jake continued. "When I was researching it, I learned that Victorian stages were actually 4x5. Their aspect ratio was the same as old TVs and computer monitors. But I wanted to build my Victorian theater in 16x9 like new TVs, computers, and movies. But it's a theater. Some of the inspiration came from the stage design from the show Wicked. It was a mix of steampunk and art deco as well."

"The frame around the monitor is a knickknack shelf that I found at the dump," Jake explained. "I trimmed the shelf to fit the size of that screen. You'll notice down at the bottom, there's some red velour." James and Brian leaned into their computer screen, which in turn displayed the image of Jake's much more beautiful computer screen. "That's intended to suggest seats. I wanted it to look like red velour theater seats."

"That's awesome," Brian called out. "I hadn't noticed that before."

"Yeah," Jake smiled. "And on either side of the screen there's fabric behind those brass grills. That's meant to suggest the theater curtains. Initially I had intended them to be actual curtains. I wanted them to run on a little servo motor, so that when the PC went into screensaver mode or went to sleep, the little curtains on either side would physically draw across the screen."

"Oh goodness. That's amazing," Brian couldn't help himself. "How about the keyboard and the rest of the kit?"

"The keyboard is the original steampunk keyboard I made. It was my first big project." (This is a very famous keyboard; more about that shortly.) "It's an IBM model M keyboard. It was the original keyboard they shipped with the first IBM PCs. It's an incredibly long-lasting keyboard and it's known for the clickety-clack noise it makes when you type. It would drive your office mates crazy because it was so loud," Jake laughed. "I replaced the tops of the keys with keys from an old Royal typewriter and built a brass frame to hold it. If you look at the image you'll see a device where the mouse is. That's just a surveyor's transit I had kicking around. When I staged the photograph, I realized that I didn't have any appropriate steampunk mice. So I grabbed that and stuck a shoelace under it to look like a cord." Jake laughed again. "That's a complete red herring there."

"Wow," James pointed at the apparatus. "This is even better than I thought. It just about covers all the different parts of steampunk culture: from modding and making, digging deep into the history—even the keyboard you used as the base for your mod has its own story—and that mouse is a lovely bit of theater in itself." James, ever the historian, had to ask: "When did you do the keyboard? That keyboard has become one of the most iconic images in steampunk."

"July 8th, 2009," Jake answered right away. "I'm looking at my website and it says the date of creation is July 8th, 2009."

"Where's the keyboard now?" Brian asked.

"It's hanging on my wall. It's a little worse for wear," Jake continued. "It's been to a lot of cons and a lot of people have pounded on it. It really needs to be freshened up."

"What was the first steampunk object you made?" Brian asked. "What inspired you to do it?"

"I guess you could say that I started doing work that was technically steampunk in 2007," Jake explained. "But really, I have always been doing stuff in this kind of style. I did a Victorian RV project in January of 2005, but no one really called it steampunk. No one really knew what that was. We just knew it was in the Victorian style and was a little bit science fiction as well. You can see the interior of the RV on my website (*http://vonslatt.com*)."

"All of the people we've talked to who have been doing steampunk for a while," James started, "the ones that everyone one knows—they've all told us the same thing. They say 'I've always done this and I only recently learned it was called steampunk—usually around 2005 to 2007.'"

"You see it in popular culture. It comes up again and again," Jake agreed. "When people talk about steampunk they make reference to all kinds of things that have come before such as the 1960s TV shows (*The Wild Wild West*) and the 1950s Disney Movies (*20,000 Leagues Under the Sea*). I think people really started recognizing it together around that time. And they did it on the Internet. It's like an echo chamber."

"It's been there for a long time, but people have only recently started to tap into it," James added.

"James's work really make me see that." Brian pointed at his exhausted cohort, who looked a little confused. "James sensitized me to culture change and how it actually happens."

"I did?" James smiled.

"Yep." Brian ignored him. "From a culture perspective, steampunk is an indicator of something going on underneath mass culture, but it's bubbling up. It's just like the hippies and the Beats getting popular. It's like when you saw someone wearing a tie-dye t-shirt in the '60s and '70s. It didn't mean they were a hippie or that they bought into the entire massive culture change that the hippies were looking for. But it meant they were interested. They were interested in that new thing which was different than what everyone else had been doing."

"I said that?" James pretended to look shocked. "That sounds pretty good."

"Goggles are the new tie-dye t-shirt. When you see goggles and top hats and all the trappings of steampunk, think culture change." Brian paused. "For me as a futurist, I see steampunk physically battling out a new relationship with technology. It's physical."

"Yeah, I think that's all true," Jake replied. "There have always been pockets of makers and those people will always be doing that but I do think they have tapped into something. I mean, they have always tapped into something but now it's getting popular."

"We wanted to ask you about the big three," James started.

"The big wha..." the phone line dropped out. "The big what?" Jake asked again.

"The big things we learned from the Vintage Tomorrows project," James explained. "We learned a lot of things but much of what we learned can be boiled down to three main areas. We learned that people's relationship with technology is changing and that people want their technology to have a sense of humor, a sense of history, and a sense of humanity."

"The first one, humor, you've hit right on!" Brian interrupted.

"The curtain," James said to Brian.

Brian nodded back and said, "Your curtain idea for the all-in-one PC is a perfect example of humor. I mean, come on! You wanted the curtain to close when the computer went to sleep. That's awesome!"

"There's a sense of wit and whimsy," James added, trying to calm Brian down. "This kind of humor is embedded in steampunk. It's really intelligent. It's like steampunk allows people to be clever."

"With me I'm not sure," Jake replied after thinking for a moment. "It might be accidental with me. Well, maybe occasionally intentional. I'm not sure. I mean, my computer mods were all built out of a sense of play. The computer has become such an important part of my life that I want it to mean more. I don't want just a piece of glass or a beige box. I want it to be something *more.* Part of that I guess could be humor. But I also think people like the feel of wood and metal. I think we all want our technology to be more tactile. We want to remember a time when our lives weren't made of plastic."

"That's what Justin Bieber said!" Brian erupted.

Jake: "Huh?"

James: "Huh?"

"Well, not Justin himself, but his manager." Brian smiled. "So you remember James's motto: 'Keep an eye out for goggles and top hats, then think culture change'? Well, in November 2011, Justin Bieber made a steampunk video for his song *Santa Claus is Coming to Town.* I saw it and thought: Culture change! So I called Scooter, his manager, and he told me that they had picked the steampunk look because it made people comfortable with new technology and it fit the feeling of the film. The song was used for a film whose story was about the past and the future coming together. Scooter said that people were attracted to the steampunk look because it made people feel more comfortable with the future. We wrote about it in the book."

"Really?" Jake said.

"Yes, Brian called Justin Bieber's people," James chuckled. "He does that."

"That makes a lot of sense," Jake replied. "You can look at vintage technology and you know how it works. It feels good to know how it works. It has a sense of history. That's something that has always been important to me."

"Your all-in-one computer one has a deep connection to the past," Brian said, pointing to the computer screen. "It's lovely, and it's lovely because you put so much thought and research into it."

"And the keyboard," James jumped in. "You modified it to look like it came straight out of the 19th century. You gave it history and a new story. You modded it into a new past."

"But it's not nostalgia," Jake wanted to be understood. "I have no desire to go back. I don't want to preserve the past at all. I'm interested in technology. I have a pocket watch that was passed down to me from my great-great-grand uncle. He was a conductor on the railroad. I don't use it. It sits in the drawer because it's a big heavy thing that only tells the time and nothing else. I have no desire for it as an object. I carry my phone; it tells time and does a host of other things, too."

"That's what we heard throughout this project," Brian agreed. "People aren't nostalgic. We know that actually the past really sucked for a lot of people. We don't want to go back there. People think that technology is cool, they just want something more from it."

"There are a lot of people using steampunk to talk about how times sucked in the past," Jake added. "Even how countries like China and India are basically in the 19th century now. You guys talked to Jaymee Goh and Magpie Killjoy right?"

"Oh yeah," Brian replied. (In Chapter 8 and the *Historian's Notebook* (see Prologue), respectively.)

"Yeah that's something that they talk about a lot," Jake added.

"Looking forward and looking backward are the same thing," James said. "All of this culture hacking and playing with history is important because it's not only about pointing out that so much of the past sucked, but we are also designing our own futures by having these discussions about the past. History is now. History is imagination."

"I have to tell you, Jake," Brian jumped in. "As a futurist I talk all the time about science fiction and having a vision for the future. I always tell people we need to have a vision for the future. We need to say 'This is the future I want to live in.'"

"Makes sense," Jake replied.

"Midway through this project James called me out and told me that it's the same thing for history," Brian continued. "History, like the future, is always in motion. We not only need to talk about the future that we want to be a part of, but we also need to talk about the history that we want to be from. I think that's amazing and true."

"Oh that's interesting," Jake said.

"Jake, I'm not sure if you remember when we first talked about this project..." James started.

"Oh, I remember," Jake laughed.

"We were having lunch with a bunch of folks before Steamcon, and I said, 'Hi, I'm a historian and I'm working on this project about steampunk and what it can teach us about the future,' and you said, 'I'm just so sick of that crap.'"

Jake laughed as the phone cut in and out.

"But when you saw the rough cut of Byrd's documentary for *Vintage Tomorrows* your reaction to steampunk was different," James continued. "You really enjoyed it and it gave you hope for steampunk culture again. We were really flattered and moved by that response."

"I liked the movie a lot," Jake replied. "It got into things that don't usually get covered. I really liked the conversation with Magpie and Jaymee. They're using steampunk for really interesting purposes. Everybody co-opts steampunk for their own passions. For me it's the technology: historical technology and making. Jaymee uses it for social justice and education. Magpie uses it to talk to people about anarchy and government. The documentary really captured that. It captured what I think steampunk is good at and why it will be around for a while. Lots of different people can use it in different ways."

"I do have to point out the wonderful thing that just happened," Brian stopped the conversation. "This is just another example of our funny little M.C. Escher project. We are talking to the guy who gave us the cover of the book about the documentary that was made based on the book. And all before the book and the documentary have been released. Awesome!"

Everyone laughed.

"I guess that is pretty funny," Jake said.

"To finish things up," Brian kept going. "We wanted to ask you where you think this is all going. What's next?"

"I have been looking for inspiration by mashing up shapes and design elements from the art nouveau movement and from early 20th century high-energy physics laboratories. A lot of the early high-energy physics stuff was in the Bugs Bunny cartoons with Marvin the Martian and also *The Jetsons*. Those cartoons are taken straight from early 20th century particle physics labs. There's an interesting blend of that aesthetic and some art nouveau. I want to do something different than all the goddamn clockwork."

"That makes perfect sense to me as a futurist." Brian got excited. "We could call it physics-punk! But that makes sense for what's next. When I look 10 to 15 years out, I see that computing is going to disappear. It's going to get so small that we'll be able to make anything into a computer. So if steampunk is our culture coming to grips with technology and devices that are woven into the fabric of our lives, then this new world fits perfectly. How do we come to grips with the fact that we are actually living in a world where we are surrounded by computational intelligence? What will it mean to live inside a computer?"

"Yes," Jake agreed. "I think the show that gets that ubiquitous computing environment right is *Star Trek: The Next Generation*. They say things like: 'Computer, where's Commander Riker?' and the ship just replies: 'Commander Riker is on the holodeck.' That's the perfect human device. You ask the air and the answer comes back to you."

"I think we are at a very important historical moment," James started. "I really saw this when I was down in Key West with William Gibson, Margret Atwood, James Gleick, and a bunch of other writers who were definitely not steampunk. When I was talking to them it became obvious that we are at a point in our culture when we know that the stories we tell ourselves about the future really *matter*. They matter so much that they can actually shape our future. Right now—today—we are writing and creating and making our future."

"So you've discovered our master plan." Jake laughed loudly for a little too long. "But of course I can't let you both live now..."

What's Next?

The project is done! All that's left is the editing both for the documentary (thanks Alan!) and the book (thanks Courtney and Brian!). As James gets ready to leave the next day for Burning Man and Brian flies out of town, the boys sit down for one last drink.

Last Call

Brian David Johnson and James H. Carrott (Portland, OR)

After we finished the interview with Jake we packed up our computers and headed to the bar. It wasn't going to be a late night. Not like the one that started this whole book/movie project. James had to get up early and continue south towards the Playa and Burning Man. Brian had to get up at 3AM to catch an early flight out for a speaking engagement. The boys would have to keep it short.

The bar at the Governor hotel in Portland, Oregon was just what you'd expect from a century-old hotel. A long heavy bar curved in the center of a richly wood-paneled room. Small tables and chairs dotted the scuffed and well-traveled floor. Off at the back was the fancier restaurant. You could tell by the bright white table linens and sparkling wine glasses.

James and Brian grabbed a table near the bar. It was early Friday afternoon and the bar was nearly full with business people grabbing a drink before dinner. Tourists stormed through the bar, their hands full of massive shopping bags. More than a few times they slapped Brian in the head with their bags as they wandered by.

After ordering two pints of IPA and two shots of whiskey James remarked, "Clearly this has become our drink order of choice for the book." James smiled. "I told Jepson if he wants to go drinking with us he needs to drink the whiskey. I don't think he likes whiskey."

Jepson referred to Brian Jepson—one of our editors for this book. He was crazy enough, along with the good people at O'Reilly, to not only agree to publish the story of our journey through steampunk and into the future of technology; they actually encouraged us to make it longer; more stories from the road; more conversations with people. "I think that sounds great," Jepson would always say in his easy voice. He's a maker. Frequently he'd tell us that he had to take a break from editing the book so he could go weld a blast plate to something. We didn't know what a blast plate was nor why you would need to weld it to anything but we were too afraid to ask him what he was talking about.

The beers and whiskey arrived and we toasted to Henry Jenkins. We had just received his foreword to the book and we were both still blown away. Henry is awesome.

"Talking to Jake got me thinking," Brian started. "The end of the interview where you asked what's next. I think he's right. I think we might know what comes after steampunk."

"Really?" James was skeptical.

"Yeah. I think we can look at the technology and that gives us some hints. "

"Of course you would say that." James smirked. "You're a technologist and a principal engineer."

"True." Brian took his point. "But hear me out, because I think we can actually see what might be coming next. "

Compute Moves to Zero

One of the most remarkable things that is heading our way is both awesome and really, really tiny. When we talk with silicon architects and engineers we see that as we head to the year 2020, the size of meaningful computational power approaches zero. This is the size of the chip by volume. We are talking about sizes that are around 14 to 7 nanometers. That's smaller than the size of most bacteria, which are typically about 20 nanometers in size.

This changes everything. For decades in computers and device manufacturing, we asked ourselves *can* we do something. Can we make a big computer workstation small enough to fit on your desk? Can we make a desktop computer small enough to fit on your lap! Can we make a laptop small enough to fit in your pocket? But when compute moves to zero all that changes. No longer do we have to ask ourselves can we do something. Now the question is *what*. What do we want to do?

That's what's next. We need to work through that as a culture and get an idea of what we want to do when we are living in a world where anything can become a computer. Your chair, your glasses, the lamps, even your own body. What do you want to do when you are living in a world where you are surrounded by intelligence?

Steampunk is the bold and wonderful exploration of what it means to have technology knit deeply into our lives. Steampunk is playing out how we want something different from our technologies. It's a device-driven exploration. The steampunk and the makers and hackers are making stuff. And they want that stuff to be more human. Not separate, but funny, with a history, and *human*.

But when we are living in a world where computers and devices aren't computers and devices anymore, how we act and interact with that intelligence needs to evolve. If you can turn your shirt into a smart phone then how do you make a phone call or send a text or even post to your favorite social networking site?

It's about relationships. Part of this idea came from our conversations with Dr. Genevieve Bell. When we live in a world of computational intelligence, then how we interact with that intelligence may not be through a keyboard or touch or even voice control. What if you interacted with these new devices by just *living* with them?

In geek terms this is called I/O. That stands for input and output. How do you get information into the computer and how does that computer give you information back? Think a keyboard (input) and a video monitor (output) or the touch screen of your smartphone (input) and your headphones for listening to music (output).

But in this new world what if the input and output is a relationship? What if you develop a relationship with all these devices and they all work together to make your life better? To make you healthier; to connect you to the people you love and the entertainment and stories you can't live without?

OK that's cool but how do we do it? Where do we start?

That's what's next. Just like Jake was thinking about the Jetsons and Star Trek —he was searching for the new metaphors for us to understand where we are going. Steampunk uses gears and pomp to express our changing relationship with technology. What is the kit of this new world? What does that look like?

The answer is easy. Keep an eye on the makers and the hackers, the builders and the artists. Now that we have sensitized you to the delicious link between cultural change and technology, you'll see it. Just like you can recognize the tie dye t-shirts or the goggles and top hats as indicators of culture change... Now you can keep an eye out for the next thing. Just keep looking. You'll know it when you see it. And it's going to be awesome.

Image Credits

A number of wonderful people and organizations have kindly granted us permissions for the illustrations that adorn this book (and its companion, the *Historian's Notebook* (see Prologue)). We'd like to call them out here for special thanks and to encourage you to check out their other work.

- Dmitri Arbacauskas (*http://www.tormentedartifacts.com/*)
- Libby Bulloff (*http://exoskeletoncabaret.com/*)
- Greg Broadmore (*http://www.gregbroadmore.com/*)
- Black Rock City, LLC (*http://www.burningman.com/*)
- Combusion Books (*http://www.combustionbooks.org/*)
- Samuel Coniglio (*http://www.samsphotography.net/*)
- Molly Michelle Friedrich (*http://porkshanks.deviantart.com/*)
- Paul Guinan (*http://www.bigredhair.com/boilerplate/*)
- Claire Hummel (*http://www.shoomlah.com/*)
- Key West Literary Seminar (*http://www.kwls.org/*) (special thanks to Arlo Haskell)
- Lionhead Studios (*http://lionhead.com/*) (special thanks to Patrick Perkins)
- David Maliki ! (*http://wondermark.com/*)
- Obtanium Works (*http://www.obtainiumworks.net/*)
- Andy Pischalnikoff (*http://www.playarazzi.com/*)
- Porter Panther (*http://porterpanther.com/*) (special thanks to Alan Winston and Byrd McDonald)
- Kimric Smythe (*http://www.etsy.com/people/NEVERWAS*)

- *SteamPunk Magazine* (*http://www.steampunkmagazine.com/*) (special thanks to Margaret Killjoy)
- Josh and Karen Tanenbaum (*http://tf.thegeekmovement.com/wp/*)
- Mark Thomson (*http://www.ibys.org/shed/*)
- Jake von Slatt (*http://steampunkworkshop.com/*)
- Weta Workshop (*http://www.wetanz.com/*)

Index

Symbols

A

B

We'd like to hear your suggestions for improving our indexes. Send email to index@oreilly.com.

D

E

I

J

T

X

Y

Z

About the Authors

James H. Carrott may have been born a historian, but definitive proof awaits further mapping of the human genome. A self-described tech nerd, anachronist, game geek, fanboy, and contrarian, James has followed an eclectic career path that has taken him from the deepest recesses of America's colonial past to the future of gaming and entertainment and everywhen between. Among many other things, he's been a miniature strategy game national champion, co-founder of a community radio station, union steward and treasurer, host and producer of innumerable radio programs, and once had the San Francisco Mime Troupe over for supper. Prior to embarking on his Vintage Tomorrows adventure, he served as global product manager for Xbox 360 hardware. James (aka CultHistorian) is currently a freelance historian, writer, and design consultant, researching cultural change to explore the future through the creative application of the past. He resides in Seattle, Washington with his two daughters in a little flat packed with books, comics, games, and toys.

The future is Brian David Johnson's business. As a futurist at Intel Corporation, his charter is to develop an actionable vision for computing in 2020. His work is called "future casting"—using ethnographic field studies, technology research, trend data, and even science fiction to provide Intel with a pragmatic vision of consumers and computing. Along with reinventing TV, Johnson has been pioneering development in artificial intelligence, robotics, and using science fiction as a design tool. He speaks and writes extensively about future technologies in articles and scientific papers as well as science fiction short stories and novels (*Fake Plastic Love, Nebulous Mechanisms: The Dr. Simon Egerton Stories* and the forthcoming *This Is Planet Earth*). He has directed two feature films and is an illustrator and commissioned painter.

Colophon

The text font is ScalaPro, designed by Martin Majoor. The heading fonts are Benton Sans and Glypha—the former font was an adaptation by Tobias Frere-Jones from News Goth, which in turn was designed by Morris Fuller Benton, and the latter font was designed by Adrian Frutiger.

CPSIA information can be obtained at www.ICGtesting.com
Printed in the USA
BVOW010604310113

312007BV00003BD/4/P

9 781449 337995